Sustainable Aviation

Series Editors

The Sustainable Aviation book series focuses on sustainability in aviation, considering all aspects of the field. The books are developed in partnership with the International Sustainable Aviation Research Society (SARES). They include contributed volumes comprising select contributions to international symposiums and conferences, monographs, and professional books focused on all aspects of sustainable aviation. The series aims at publishing state-of-the-art research and development in areas including, but not limited to:

- Green and renewable energy resources and aviation technologies
- Aircraft engine, control systems, production, storage, efficiency, and planning
- Exploring the potential of integrating renewables within airports
- Sustainable infrastructure development under a changing climate
- Training and awareness facilities with aviation sector and social levels
- Teaching and professional development in renewable energy technologies and sustainability

* * *

T. Hikmet Karakoc • Tomislav Letnik
Maršenka Marksel • Ismail Ekmekci
Alper Dalkiran • Ali Haydar Ercan
Editors

Emerging Trends in Electric Aviation

Proceedings of the International Symposium on Electric Aviation and Autonomous Systems 2022

Springer

Editors
T. Hikmet Karakoc
Faculty of Aeronautics and Astronautics
Eskisehir Technical University
Eskisehir, Türkiye

Information Technology Research
and Application Center
Istanbul Ticaret University
Istanbul, Türkiye

Maršenka Marksel
FPGA
University of Maribor
Maribor, Slovenia

Alper Dalkiran
School of Aviation
Süleyman Demirel University
Keciborlu, Isparta, Türkiye

Tomislav Letnik
FGPA
University of Maribor
Maribor, Slovenia

Ismail Ekmekci
Faculty of Engineering
Istanbul Commerce University
Istanbul, Türkiye

Ali Haydar Ercan
Porsuk Vocational School
Eskisehir Technical University
Eskisehir, Türkiye

ISSN 2730-7778 ISSN 2730-7786 (electronic)
Sustainable Aviation
ISBN 978-3-031-37298-8 ISBN 978-3-031-37299-5 (eBook)
https://doi.org/10.1007/978-3-031-37299-5

This Springer imprint is published by the registered company Springer Nature Switzerland AG
The registered company address is: Gewerbestrasse 11, 6330 Cham, Switzerland

Preface

New fuel and propulsion technologies are being proposed to find suitable alternatives to fossil fuels and new green energy production methods, such as battery-powered and hydrogen-powered aircraft, hybrid systems, and synthetic fuels. Various processes can produce hydrogen, but achieving a low maximum take-off mass is the biggest challenge. Fuel cell aircraft have a low power-to-weight ratio, and the cost of depreciation, interest, and insurance is proportional to the price of the aircraft; also, a viable and environmentally friendly alternative to conventional aircraft, with safety considerations, regulatory issues, ground infrastructure, and government and public acceptance.

On the other hand, additive manufacturing, an umbrella term for different manufacturing methods, aims to manufacture complex three-dimensional shapes by adding materials successively and their sequential connection to each other. This technique may help to reduce and overcome power over weight problems of future aspects of electric aircraft. Topological optimization is a new technology that optimizes material layout within a given design space for a given set of loads, boundary conditions, and constraints. Additive manufacturing enables solutions to increase efficiency and reduce parts weight, but certification is challenging.

Nevertheless, it seems there will not be an alternative energy resource for fully electric aircraft for a midterm period than using the batteries. Moreover, resolutions on battery charging performance are moving forward. One of the performance issues for battery energy is discharging and charging. In order to fully discharge the batteries of an electric aircraft at minimum cruise power and examining the motor power periodically as well as battery temperature may become necessary on operations. Electric aircraft performance changes during a flight are different than what a pilot expects from a gasoline-powered aircraft, resulting in increased performance near the end of a flight but lower performance at lower battery SOCs.

Hybrid Electric Propulsion solutions are a viable turnaround for reducing emissions in general aviation by decoupling the Internal Combustion Engine from the propulsor to the electric aircraft and allowing for a combustion engine downsizing and operation in the best economy point. A serial hybrid propulsion system has been implemented into a modern general aviation aircraft to demonstrate the possibility of a low fuel consumption two-seater. However, those solutions haven't been enlightened a revolutionary transition over power to weight dilemma.

Electrical aircraft's comparatively low energy storage capacity is the only serious obstacle to developing successful regional aircraft fleets and achieving zero-emission flight. The next-generation air passenger experience will begin with the development of these two major performance issues.

ISEAS '22, an international and multi-disciplinary symposium on electric aviation and autonomous solutions, was held online between July 19 and 21, 2022, to address electric aircraft systems and safe, reliable electric power in aviation. We have kindly invited academics, scientists, engineers, practitioners, policymakers, and students to attend the ISEAS symposium to share knowledge, explain new technologies and discoveries, and consider the future direction, strategies, and goals in maintenance. This conference featured keynote presentations by invited speakers and general papers in oral and poster sessions.

We want to thank Springer's editorial team for their support toward the preparation of this book and the chapter authors and reviewers for their outstanding efforts.

We would also like to give special thanks to the SARES Editorial office members for gathering these chapters, who are the heroes behind the veil of the stage. Dilara Kılıc played a significant role in sharing the load and managing the chapters with Sinem Can. Also, we thank Kemal Keles for his efforts in the long run for symposium author communications.

Eskisehir, Turkiye — T. Hikmet Karakoc
Maribor, Slovenia — Tomislav Letnik
Maribor, Slovenia — Maršenka Marksel
Istanbul, Turkiye — Ismail Ekmekci
Keciborlu, Turkiye — Alper Dalkiran
Eskisehir, Turkiye — Ali Haydar Ercan

Contents

About the Editors

Alper Dalkıran received his bachelor's degree from the Avionics Department, Faculty of Aeronautics and Astronautics, Eskisehir Technical University (formerly known as Anadolu University). He completed his M.Sc. degree at the School of Science, Anadolu University in 2004 in Aviation Maintenance. He earned his Ph.D. degree in 2017 in Environmental Sustainability on Airports from the School of Science, Anadolu University by developing a model of energy-based calculations of an aerodrome. He has studied on aircraft engines, sustainability, airports, and exergy. He has 17 years of professional experience in Airports in Information Technology, Automation, and Integration. He has managed teams on system design, projects, tests, commissioning, operational readiness, and operations. He has been working in the School of Aviation at Suleyman Demirel University since 2019 and lecturing in Flight Theory, Airline Management, and Airport Design subjects.

Ali Haydar Ercan received his bachelor's degree from the Mechanical Engineering Department, Faculty of Engineering, University of Cumhuriyet, Sivas, and his M.Sc. degree from the Mechanical Engineering Department, University of Gazi, Ankara, where he studied Heat Transfer. He earned his Ph.D. from the Department of Aerodynamics, University of Liverpool, England, where he worked on Boundary Layer Theory on Flat Plate Surfaces and developed empiric formulas for transition development distance from the leading edge. He completed his Ph.D. in 1997. He also earned a postgraduate degree in Software Technologies from the University of Liverpool in 1999. He worked at the University of Cumhuriyet as a Lecturer, at the University of Liverpool as an Assistant Lecturer while studying for his Ph.D., at the University of 19 Mayıs as Head of the Junior Aviation Technical School, and was one of the founders of the department. Currently, he is working as a Lecturer at the

Department of Unmanned Aerial Vehicles in Electronics and Automation, Porsuk Vocational School, Eskisehir Technical University. He is also working on PIV (Particle Image Velocimetry)-related subjects and a new idea for wing design. He is a reviewer for four journals and a guest editor. He has held a wide range of managerial positions experience with international commercial private companies.

Ismail Ekmeki is a Lecturer at the Department of Industrial Engineering, Istanbul Commerce University. He received his M.Sc. in Mechanical Engineering from the School of Science at Yildiz Technical University, an M.Sc. in Industrial Engineering from İstanbul Technical University, and a Ph.D. in Mechanical Engineering from Yildiz Technical University in 1982, 1983, and 1995, respectively. He has worked as a researcher, executive, and manager in academic positions on various industrial projects. His areas of research cover energy, optimization, decision making, safety, OHS, and HVAC-related topics.

T. Hikmet Karakoc graduated from the Department of Mechanical Engineering, Anadolu University. He received his M.Sc. degree in Mechanical Engineering from the Yildiz Technical University. He received his Ph.D. from Anadolu University, where he started his full-time teaching and received his Full Professorship. He is currently researching at the Eskisehir Technical University. He has a wide range of research interests, including Sustainable Aviation, Aircraft Propulsion Systems, Insulation, Heating, Ventilating, and Air Conditioning, Indoor Air Quality, Gas Turbines, Cogeneration Systems, Renewable Energy, Energy Economics, Fuels, and Combustion. He has participated in numerous industrial projects on these topics as a researcher, consultant, and project manager for over 30 projects and corporations. He also started a contest on special insulation applications among university students. He served as an editor-in-chief, guest editor, and editorial board member for international scientific journals. He published national and international papers in over 300 journals and 40 books. Professor Karakoc actively follows membership positions for the Chamber of Mechanical Engineers and many sectorial associations, international scientific organizations, and societies. He is an active Board of Directors member of the International Association for Green Energy. He is currently holding the presidency of the SARES organization, which is actively supporting scientists and students in the area of Sustainable Aviation. He also organizes four symposiums on aviation subject areas as a Founding Chair.

Tomislav Letnik is Head of the Transport Economics Centre at the University of Maribor, where he has been a project manager since 2005. His work involves the management of several European and national projects in the field of logistics and transport economics. Dr. Letnik is also a Lecturer at the University of Maribor for courses on logistics, transport economics, investment decisions, and system theory. He holds a B.S. degree in Transport Engineering and has successfully concluded postgraduate studies at the University of Maribor.

Maršenka Marksel is a Research Assistant at the Faculty of Civil Engineering, Transportation Engineering, and Architecture, University of Maribor. She received a Bachelor's degree from the University of Maribor Faculty of Economics and Business. During her studies, she was actively involved in several research projects at the Institute of Economic Analysis and Forecasting and the Transport Economic Centre at the University of Maribor.

Gas Turbine and Fuel Cell Hybrid Systems

Enes Gunaltili, Selcuk Ekici, Mustafa Zeki Yilmazoglu, and Tahir Hikmet Karakoc

Nomenclature

DAFC	Direct Alcohol Fuel Cells
DBFC	Direct Borohydride Fuel Cell
MCFC	Molten Carbonate Fuel Cells
PAFC	Phosphoric Acid Fuel Cells
PEMFC	Proton Exchange Membrane Fuel Cells
SOFC	Solid Oxide Fuel Cells

1 Introduction

The use of conventional energy sources (fossil fuels) in gas turbine systems has various impacts on both the global and local atmosphere. These impacts are local air pollution triggered by greenhouse gas emissions (Santibanez-Borda et al. 2021). In this context, the importance of using renewable and alternative energy sources (e.g., biofuels, solar energy, wind energy, and fuel cells), especially in gas turbine engines

E. Gunaltili (✉)
Faculty of Aeronautics and Astronautics, Necmettin Erbakan University, Konya, Türkiye

S. Ekici
Faculty of Economics and Administrative Sciences, Igdir University, Igdir, Türkiye

M. Z. Yilmazoglu
Faculty of Engineering, Gazi University, Ankara, Türkiye

T. H. Karakoc
Faculty of Aeronautics and Astronautics, Eskişehir Technical University, Eskisehir, Türkiye

Information Technology Research and Application Center, Istanbul Ticaret University, Istanbul, Türkiye

T. H. Karakoc et al. (eds.), *Emerging Trends in Electric Aviation*, Sustainable Aviation, https://doi.org/10.1007/978-3-031-37299-5_1

that consume fuel at high mass flow rates, is increasing day by day in terms of both the atmosphere and human health (Lefebvre and Ballal 2010). Today, many studies have been conducted, and methods have been developed for the hybridization of gas turbine-fuel cell hybrid systems (Buonomano et al. 2015). Especially since the studies on the adaptation of hybridization methods used in industrial hybrid systems to gas turbine-fuel cell hybrid systems in the aviation field are quite new, there is a need for a study in this field in the literature (Fernandes et al. 2018). Hybridization systems (Seitz et al. 2022) can be classified according to their characteristics, such as fuel cell type, cycle type, cycle operating pressure, reformer (Zhang et al. 2014) and steam reforming type, and circulation type. The use of aviation gas turbine-fuel cell hybrid systems as auxiliary power units in large-scale aircraft (Arat et al. 2020) or as the main power unit in unmanned aerial vehicles (Zhixing et al. 2020) is increasing day by day. For this reason, mathematical modeling and optimization of such hybrid system studies have been studied with various modeling and simulation techniques, which have become very popular recently (Ameri and Mohammadi 2013).

2 Gas Turbine Fuel Cell Hybrid Systems

In general, gas turbine engines are classified as turbojet, turbofan, turboshaft, or turboprop according to their intended use, the method of channeling the air, and the way the thrust is generated through the engine flow paths (El-Sayed 2017). Turbojet engines are mostly used in military aircraft today due to their advantages, such as their high thrust generation capacity and ability to reach high speeds, as well as their disadvantages, such as their noisy operation, short range, high fuel consumption, and inefficient operation at low speeds (Benini and Giacometti 2007). Turbofan engines, on the other hand, are widely used in passenger aircraft, designed for commercial airline service due to their advantages such as low fuel consumption, quiet operation, and long-range (Richter 2012). Turboprop engines are mostly preferred for heavy cargo aircraft due to their advantages, such as low emissions, the ability to operate at a wide range of speeds, and high efficiency at different altitudes, but also their disadvantages, such as high fuel consumption and a short range (Jagadish Babu et al. 2019). Turboshaft engines, another type of gas turbine engine, are generally preferred in helicopters due to their noisy operating conditions and complex structure (Misté and Benini 2012). In the design, modeling, and optimizations to be carried out for all mentioned gas turbine models, operating conditions must be taken into consideration. In engines operating at high temperatures, attention should be paid to the correct material selection and regular maintenance, especially in terms of engine life (El-Sayed 2017).

In today's world, where the demand for renewable energy sources is increasing, fuel cells are an important power source because they are highly efficient, clean, and sustainable energy sources. In general, fuel cells can be divided into several different categories according to the types of electrolytes and fuel used (Singh et al. 2021). Each fuel cell type has its own advantages and disadvantages. With recent new studies and projects, fuel cells have started to be used in both power systems and transportation-based systems from an environmental, commercial, and industrial perspective (Baroutaji et al. 2019). Planar and tubular fuel cells are the two primary categories that may be classified among fuel cells according to their geometric form (Huang and Singhal 2013). The primary benefits of tubular fuel cell systems are the ease of fuel and air supply, the capacity to withstand greater thermal stresses, and the quick reaction to load fluctuation (Hatchwell et al. 1998). Because of its high power density and low production cost, planar geometry is preferred in fuel cell stacks (Singh et al. 2021). Fuel cells, which have recently become widely used due to factors such as sustainability and environmentalism, offer various advantages and disadvantages according to their types (Sazali et al. 2020).

The most common types of fuel cells may be categorized as follows:

- Proton Exchange Membrane Fuel Cells (PEMFC) (Jiao et al. 2021).
- Solid Oxide Fuel Cells (SOFC) (Peng et al. 2021).
- Molten Carbonate Fuel Cells (MCFC) (Mehr et al. 2021).
- Phosphoric Acid Fuel Cells (PAFC) (Guo et al. 2021).
- Alkaline Fuel Cells (AFC) (Ferriday and Middleton 2021).
- Direct Alcohol Fuel Cells (DAFC) (Shaari et al. 2021).
- Direct Borohydride Fuel Cell (DBFC) (Yaqoob et al. 2021).

and each fuel cell has systemic advantages and disadvantages (Sazali et al. 2020). Mathematical models that meet the purpose of the system to be used are very important for designing and optimizing fuel cell systems (Li et al. 2020). Studies on modeling various kinds of fuel cells, including Proton Exchange Membrane Fuel Cells, Solid Oxide Fuel Cells, and Molten Carbonate Fuel Cells, have recently started to be carried out. Specifically, the modeling and simulation of these three kinds of fuel cells are being done with various techniques (Rosner et al. 2020). Before beginning to model a hybrid system, it is vital to first decide what the objective of the model is. Then, depending on the objective, it is necessary to figure out what the model's main characteristics are.

The gas turbine-fuel cell hybrid cycle can be created in its simplest form in the pressurized Brayton cycle by changing the combustor with a fuel cell stack that provides heat to the mixture expanded in the turbine and generates electricity (McLarty et al. 2012). In addition, by adding fuel cells as an auxiliary unit to the system, a system plan must be used to increase system efficiency and reduce emissions. Many alternative configurations of fuel cell-gas turbine hybrid systems can be created by creating these or similar system structures with hybridization techniques (Liu et al. 2021). Today, the increasing importance of low emission and sustainability requirements, especially in aviation, and the many regulations related to this have led to the need for studies on such hybrid systems. Therefore, in the last few years, researchers have developed numerous hybrid system configurations aimed at increasing electrical efficiency and reducing system costs (Buonomano et al. 2015). Hybrid systems that use both fuel cells and gas turbines can be used for many things. Two examples are auxiliary power units in aviation and primary power units in unmanned aerial vehicles (Alrashed et al. 2021). The development of aviation technology has resulted in a large increase in the quantity of energy demand in recent years (Bachmann et al. 2017). When designing fuel cell-based auxiliary power systems, additional elements such as pumps, sensors, blowers, reformers, controllers, etc. need to be used in the hybrid system to adjust the operating system and efficiency of the fuel cell stack (Guo et al. 2020). The advantages of fuel cells, such as high specific energy, noiseless operation, high efficiency, low weight, environmentalism, and sustainability, make these systems important for the aviation industry (Azizi and Brouwer 2018). These advantages increase the use of fuel cell stacks in unmanned aerial vehicles. However, the disadvantages of fuel cells, such as their limited power density and slow dynamic response, need to be overcome with proper modeling and planning in system designs (Townsend et al. 2020). Furthermore, unmanned aerial vehicles operate in a variety of flight conditions with varying ambient temperatures and pressures, causing issues with the constant operating conditions of fuel cells. (Wilberforce et al. 2019). Therefore, special measures and procedures are also required to maintain the performance of fuel cells. In hybrid system designs, where they are preferred as the additional power system in airplanes, the power-to-weight ratio is an important parameter for the hybrid system (Zhang et al. 2021). In order to improve this ratio, it is necessary to focus on low-weight fuselage or wing structures and high-efficiency system designs (Grebenikov et al. 2020). In this regard, it should be taken into account that the additional structures that need to be used in the system in order to create hybrid operating conditions for fuel cells and to ensure efficiency increases also increase the system's weight and add complexity to the system. In addition, in order to accomplish the required performance levels, the technology needs to address issues such as low specific power, the requirement to reform aviation fuels, durability, safety, and cost (Hodjati-Pugh et al. 2021).

3 Conclusion

The efficiency of the hybrid fuel cell-gas turbine system is dependent on the individual performance characteristics of the gas turbines and fuel cells. While pressure ratios, mass flow rates, turbine inlet temperature, shaft speed, and efficiency are the main parameters for gas turbines (Kurz and Brun 2016), the main parameters for fuel cell operation are power, pressure, voltage current, efficiency, temperature, and fuel utilization (Santin et al. 2008). Hybrid system simulation is important since it focuses on how the fuel cell interacts with the overall system parts and how that might impact the system's total efficiency (Damo et al. 2019). Various assumptions and simplifications need to be made in system modeling because of the complicated framework of fuel cell-gas turbine hybrid systems. In order to achieve this goal, the models that were developed take into consideration the relationship between electrochemistry, mass transfer, chemistry, and heat transfer. This allows for the optimization of cell design and performance within the context of the fuel cell design part (Buonomano et al. 2015). The modeling in the gas turbine engine design section aims to reduce specific fuel consumption by increasing efficiency and propulsion power through the main parameters in the performance calculations of the engines (Arsalis 2019).

The research opportunities and concerns of fuel cell-gas turbine hybrid systems are generally focused on producing more efficient and greener power systems. Hybrid systems, especially those designed for industrial hybrid power systems, offer great potential for developments in the aviation sector. The issues that need to be considered here are modeling in accordance with the operating conditions and system dynamic responses for unmanned aerial vehicle systems and planning the weight impact of the parts used in hybrid systems designed as auxiliary power units for large-scale aircraft. In future studies, the systems to be designed with next-generation hybridization techniques should be developed by paying attention to the power-to-weight ratio. Focusing on studies in this field will enable the development of propulsion systems for the aviation industry, which has a very large market share. For this reason, it is clear that more research and development is needed on fuel cell-gas turbine hybrid systems.

References

Alrashed M, Nikolaidis T, Pilidis P, Jafari S (2021) Utilisation of turboelectric distribution propulsion in commercial aviation: a review on NASA's TeDP concept. Chin J Aeronaut 34: 48–65. https://doi.org/10.1016/j.cja.2021.03.014

Ameri M, Mohammadi R (2013) Simulation of an atmospheric SOFC and gas turbine hybrid system using Aspen plus software. Int J Energy Res 37:412–425. https://doi.org/10.1002/er.1941

Arat HT, Sürer MG, Gökpinar S, Aydin K (2020) Conceptual design analysis for a lightweight aircraft with a fuel cell hybrid propulsion system. Energy Sources A: Recovery Util Environ Effects:1–15. https://doi.org/10.1080/15567036.2020.1773966

Arsalis A (2019) A comprehensive review of fuel cell-based micro-combined-heat-and-power systems. Renew Sust Energ Rev 105:391–414. https://doi.org/10.1016/j.rser.2019.02.013

Azizi MA, Brouwer J (2018) Progress in solid oxide fuel cell-gas turbine hybrid power systems: system design and analysis, transient operation, controls and optimization. Appl Energy 215: 237–289. https://doi.org/10.1016/j.apenergy.2018.01.098

Bachmann J, Hidalgo C, Bricout S (2017) Environmental analysis of innovative sustainable composites with potential use in aviation sector – a life cycle assessment review. Sci China Technol Sci 60:1301–1317. https://doi.org/10.1007/s11431-016-9094-y

Baroutaji A, Wilberforce T, Ramadan M, Olabi AG (2019) Comprehensive investigation on hydrogen and fuel cell technology in the aviation and aerospace sectors. Renew Sust Energ Rev 106:31–40. https://doi.org/10.1016/j.rser.2019.02.022

Benini E, Giacometti S (2007) Design, manufacturing and operation of a small turbojet-engine for research purposes. Appl Energy 84:1102–1116. https://doi.org/10.1016/j.apenergy.2007.05.006

Buonomano A, Calise F, d'Accadia MD, Palombo A, Vicidomini M (2015) Hybrid solid oxide fuel cells–gas turbine systems for combined heat and power: a review. Appl Energy 156:32–85. https://doi.org/10.1016/j.apenergy.2015.06.027

Damo UM, Ferrari ML, Turan A, Massardo AF (2019) Solid oxide fuel cell hybrid system: a detailed review of an environmentally clean and efficient source of energy. Energy 168:235–246. https://doi.org/10.1016/j.energy.2018.11.091

El-Sayed A (2017) Aircraft propulsion and gas turbine engines, 2nd edn. CRC Press

Fernandes MD, Andrade SD, Bistritzki VN, Fonseca RM, Zacarias LG, Gonçalves HN, De Castro AF, Domingues RZ, Matencio T (2018) SOFC-APU systems for aircraft: a review. Int J Hydrog Energy 43:16311–16333. https://doi.org/10.1016/j.ijhydene.2018.07.004

Ferriday TB, Middleton PH (2021) Alkaline fuel cell technology – a review. Int J Hydrog Energy 46:18489–18510. https://doi.org/10.1016/j.ijhydene.2021.02.203

Grebenikov AG, Gumenniy AM, Buival LY, Chumak AS, Sobolev AA (2020) Light civil turboprop airplane take-off weight preliminary design estimation method. In: Nechyporuk M, Pavlikov V, Kritskiy D (eds) Integrated computer technologies in mechanical engineering, vol 1113. Springer International Publishing, Cham, pp 60–74

Guo Y, Yu Z, Li G, Zhao H (2020) Performance assessment and optimization of an integrated solid oxide fuel cell-gas turbine cogeneration system. Int J Hydrog Energy 45:17702–17716. https://doi.org/10.1016/j.ijhydene.2020.04.210

Guo F, Qin J, Ji Z, Liu H, Cheng K, Zhang S (2021) Performance analysis of a turbofan engine integrated with solid oxide fuel cells based on Al-H2O hydrogen production for more electric long-endurance UAVs. Energy Convers Manag 235:113999 . https://doi.org/10.1016/j.enconman.2021.113999

Hatchwell CE, Sammes NM, Kendall K (1998) Cathode current-collectors for a novel tubular SOFC design. J Power Sources 70:85–90. https://doi.org/10.1016/S0378-7753(97)02693-1

Hodjati-Pugh O, Dhir A, Steinberger-Wilckens R (2021) The development of current collection in micro-tubular solid oxide fuel cells – a review. Appl Sci 11:1077. https://doi.org/10.3390/app11031077

Huang K, Singhal SC (2013) Cathode-supported tubular solid oxide fuel cell technology: a critical review. J Power Sources 237:84–97. https://doi.org/10.1016/j.jpowsour.2013.03.001

Jagadish Babu C, Arul Kumaresan D, Kumar V, Ragupathy R, Mishra RK (2019) Analysis and prevention of failures in a turboprop engine. J Fail Anal Preven 19:1195–1201. https://doi.org/10.1007/s11668-019-00727-6

Jiao K, Xuan J, Du Q, Bao Z, Xie B, Wang B, Zhao Y, Fan L, Wang H, Hou Z, Huo S, Brandon NP, Yin Y, Guiver MD (2021) Designing the next generation of proton-exchange membrane fuel cells. Nature 595:361–369. https://doi.org/10.1038/s41586-021-03482-7

Kurz R, Brun K (2016) Gas turbine performance. Turbomachinery Laboratories, Texas A&M Engineering Experiment Station. Available electronically from https://hdl.handle.net/1969.1/160276

Lefebvre AH, Ballal DR (2010) Gas turbine combustion: alternative fuels and emissions, 3rd edn. CRC Press, London : Taylor & Francis [distributor], Boca Raton

Li G, Gou Y, Qiao J, Sun W, Wang Z, Sun K (2020) Recent progress of tubular solid oxide fuel cell: from materials to applications. J Power Sources 477:228693 . https://doi.org/10.1016/j.jpowsour.2020.228693

Liu H, Qin J, Ji Z, Guo F, Dong P (2021) Study on the performance comparison of three configurations of aviation fuel cell gas turbine hybrid power generation system. J Power Sources 501:230007 . https://doi.org/10.1016/j.jpowsour.2021.230007

McLarty D, Kuniba Y, Brouwer J, Samuelsen S (2012) Experimental and theoretical evidence for control requirements in solid oxide fuel cell gas turbine hybrid systems. J Power Sources 209: 195–203. https://doi.org/10.1016/j.jpowsour.2012.02.102

Mehr AS, Lanzini A, Santarelli M, Rosen MA (2021) Polygeneration systems based on high temperature fuel cell (MCFC and SOFC) technology: system design, fuel types, modeling and analysis approaches. Energy 228:120613 . https://doi.org/10.1016/j.energy.2021.120613

Misté GA, Benini E (2012) Performance of a Turboshaft engine for helicopter applications operating at variable shaft speed. In: ASME 2012 Gas Turbine India Conference. American Society of Mechanical Engineers, pp 701–715

Peng J, Huang J, Wu X, Xu Y, Chen H, Li X (2021) Solid oxide fuel cell (SOFC) performance evaluation, fault diagnosis and health control: a review. J Power Sources 505:230058 . https://doi.org/10.1016/j.jpowsour.2021.230058

Richter H (2012) Advanced control of turbofan engines. Springer, New York

Rosner F, Rao A, Samuelsen S (2020) Economics of cell design and thermal management in solid oxide fuel cells under SOFC-GT hybrid operating conditions. Energy Convers Manag 220: 112952 . https://doi.org/10.1016/j.enconman.2020.112952

Santibanez-Borda E, Korre A, Nie Z, Durucan S (2021) A multi-objective optimisation model to reduce greenhouse gas emissions and costs in offshore natural gas upstream chains. J Clean Prod 297:126625 . https://doi.org/10.1016/j.jclepro.2021.126625

Santin M, Traverso A, Massardo A (2008) Technological aspects of gas turbine and fuel cell hybrid systems for aircraft: a review. Aeronaut J 112:459–467. https://doi.org/10.1017/S0001924000002426

Sazali N, Wan Salleh WN, Jamaludin AS, Mhd Razali MN (2020) New perspectives on fuel cell technology: a brief review. Membranes (Basel) 10. https://doi.org/10.3390/membranes10050099

Seitz A, Nickl M, Troeltsch F, Ebner K (2022) Initial assessment of a fuel cell – gas turbine hybrid propulsion concept. Aerospace 9:68 . https://doi.org/10.3390/aerospace9020068

Shaari N, Kamarudin SK, Bahru R, Osman SH, Md Ishak NAI (2021) Progress and challenges: review for direct liquid fuel cell. Int J Energy Res 45:6644–6688. https://doi.org/10.1002/er.6353

Singh M, Zappa D, Comini E (2021) Solid oxide fuel cell: decade of progress, future perspectives and challenges. Int J Hydrog Energy 46:27643–27674. https://doi.org/10.1016/j.ijhydene.2021.06.020

Townsend A, Jiya IN, Martinson C, Bessarabov D, Gouws R (2020) A comprehensive review of energy sources for unmanned aerial vehicles, their shortfalls and opportunities for improvements. Heliyon 6:e05285. https://doi.org/10.1016/j.heliyon.2020.e05285

Wilberforce T, El Hassan Z, Ogungbemi E, Ijaodola O, Khatib FN, Durrant A, Thompson J, Baroutaji A, Olabi AG (2019) A comprehensive study of the effect of bipolar plate (BP) geometry design on the performance of proton exchange membrane (PEM) fuel cells. Renew Sust Energ Rev 111:236–260. https://doi.org/10.1016/j.rser.2019.04.081

Yaqoob L, Noor T, Iqbal N (2021) Recent progress in development of efficient electrocatalyst for methanol oxidation reaction in direct methanol fuel cell. Int J Energy Res 45:6550–6583. https://doi.org/10.1002/er.6316

Zhang X, Wang Y, Liu T, Chen J (2014) Theoretical basis and performance optimization analysis of a solid oxide fuel cell–gas turbine hybrid system with fuel reforming. Energy Convers Manag 86:1102–1109. https://doi.org/10.1016/j.enconman.2014.06.068

Zhang C, Zhu C, Meng B, Li S (2021) Challenges and solutions for high-speed aviation piston pumps: a review. Aerospace 8:392 . https://doi.org/10.3390/aerospace8120392

Zhixing Z, Ji Z, Qin J, Cheng K, Dang C, Zhang S, Dong P (2020) Analysis of safe operation zone for a turbine-less and solid oxide fuel cell hybrid electric jet engine on unmanned aerial vehicles

Implementation of a Two-Seat Hybrid Electric Aircraft Demonstrator for Reducing Carbon Emissions

Jonas Lay and Andreas Strohmayer

Nomenclature

HCU	Hybrid Control Unit
HEPS	Hybrid Electric Propulsion Systems
ICE	Internal Combustion Engine
MTOM	Maximum Take-Off Mass
SAF	Sustainable Aviation Fuel

1 Introduction

The reduction of emissions has become a main challenge for modern day aviation. Hybrid Electric Propulsion Systems (HEPS) could be a viable solution for the field of general aviation. In a serial HEPS, this is mainly done by decoupling the Internal Combustion Engine (ICE) from the propulsor, which no longer have coupled RPM ranges and allows for an ICE downsizing and also operation in the best economy point (Glassock et al. 2017). Other possible benefits include the reduction of noise, an increase of safety, and the lowering of operation costs. There are numerous analytical studies and concept aircraft that investigate the benefits of the application of HEPS in existing conventional aircraft and in purpose-made new designs (ICAO 2019). However, there are not many applicatory studies that focus on the real-world implications of designing and operating such a system.

The disadvantages of a usually heavier and more complex propulsion system have to be outweighed by the possible advantages in the design. As concluded in,

J. Lay (✉) · A. Strohmayer
University of Stuttgart, Stuttgart, Germany
e-mail: lay@ifb.uni-stuttgart.de; strohmayer@ifb.uni-stuttgart.de

T. H. Karakoc et al. (eds.), *Emerging Trends in Electric Aviation*, Sustainable Aviation, https://doi.org/10.1007/978-3-031-37299-5_2

light aircraft has a wider variation of power required for different flight conditions compared to large aircraft (Sziroczak et al. 2020). This can be exploited for reducing the overall fuel consumption for a given mission with a hybrid powertrain by using a thermal combustion engine in its operation point with the lowest specific fuel consumption (Donateo et al. 2017). Excess power can be stored in a battery which can be used to provide electric power during phases of high power demand.

Additionally, a serial hybrid electric system has the potential to be operated on batteries alone, which might be desirable for noise emissions especially during flight phases close to the ground, i.e., take-off and landing.

1.1 Flying Platform

Depending on the aircraft being used, different hybridization approaches are more or less beneficial when implemented. On one hand, conventional single engine tractor configurations may benefit from a parallel hybrid setup for boosting ICE power and possible engine downsizing. On the other hand, custom aircraft configurations for lower power consumption are possible with an electric propulsor only. The e-Genius is a unique example of exploiting a low drag integration for better propulsive efficiency. In that case, a serial hybrid layout can be more beneficial (Table 1).

1.2 Powertrain Layout

In Fig. 1, the general high voltage layout for the demonstrator is shown.

The system operates at a nominal voltage of 400 V. For the used power range of up to 85 kW propulsive power, this poses a good compromise between minimizing cable mass while keeping the effort that has to be taken to isolate the components low (Fig. 2).

Table 1 Technical specifications of the aircraft

Span width	16.86 m
Length	8.10 m
Wing area	14.1 m^2
Payload	195 kg
MTOM	950 kg
Shaft power (max./cont.)	85 kW/35 kW
Cruising flight	140–200 km/h
Climb rate	4 m/s
Climb rate at take-off	5.5 m/s
Take-off distance	450 m
Landing distance	350 m
Glide ratio	1:33

Fig. 1 e-Genius flying over the Alps

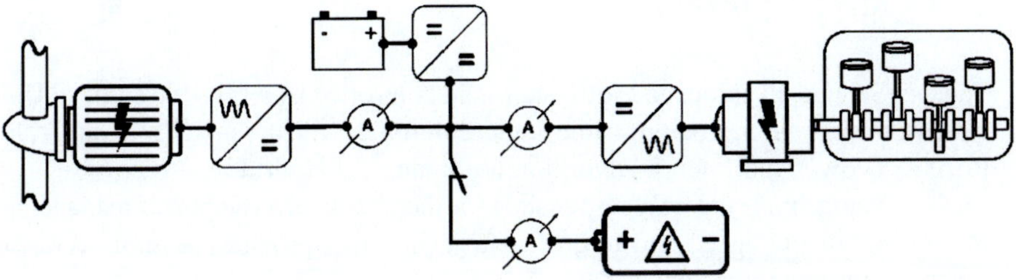

Fig. 2 Electrical layout of the propulsion chain

Each branch in the system can be deactivated by contactors or the power electronics themselves (Bravo et al. 2021). The system is designed to be able to run on batteries alone (by mission planning or in case of a generator system failure).

1.3 Generator System

The GenSet consists of a three-cylinder turbocharged multifuel engine (JET-A, Diesel, SAF) and an axial flux permanent magnet synchronous electric machine. Both are water cooled, but at different temperature levels, so two different cooling circuits have to be used, three including the charge air intercooling system.

The GenSet provides up to 40 kW, but power below 20 kW shows best efficiencies.

1.4 Integration

Integration of a serial HEPS is challenging in a small GA aircraft. The e-Genius has been specifically designed to accommodate different large experimental setups in the space behind the passengers. Especially the three mentioned cooling systems and the various fire protection measures take up substantial volume in the aircraft.

The added complexity of the propulsion system will take up more weight compared to a conventional engine or a purely electric setup. Also, the cooling systems will most likely add cooling drag. These downsides will have to be exceeded by the advantages of the more efficient thermal efficiencies in the system as well as the superior integration aspects.

With their relatively high weight, the batteries can be used to adjust the overall center of gravity in the aircraft easily.

2 Control Algorithm

The GenSet and all associated subsystems are controlled by a "Hybrid Control Unit" which consists of a programmable logic controller which runs a software called PEACH (Power and Energy Algorithm for Control of Hybrids).

This Algorithm is not only responsible for the overall electric power management in the aircraft via speed and torque control, but also provides several secondary functions in the propulsion system.

- Battery envelope protection.
- Cooling system control (flow rates, warm-up, cool-down).
- Automated GenSet start functionality.
- Fire detection and extinguisher.
- Operational parameter monitoring and protection.
- Energy monitoring (electrical capacity, fuel level, range).
- Efficiency calculation and operational assistance.
- Maintenance mod.

In Fig. 3, the signal paths of the communication are indicated. The pilot is responsible controlling the aircraft and also the main propulsor with no external systems interfering. The HCU is monitoring all systems and all necessary auxiliary sensor data, e.g., power demand and outside conditions. By the chosen flight mission, PEACH controls the power output according to the predefined operational settings. For example, in power matching mode, the output can be set to constantly match the power demand and thus can neutralize the battery current, preserving the state of charge (Fig. 4).

This way, the workload on the pilot is not increased despite significantly increased complexity and interactions in the propulsion system.

Fig. 3 Packaging of propulsion system in e-Genius fuselage

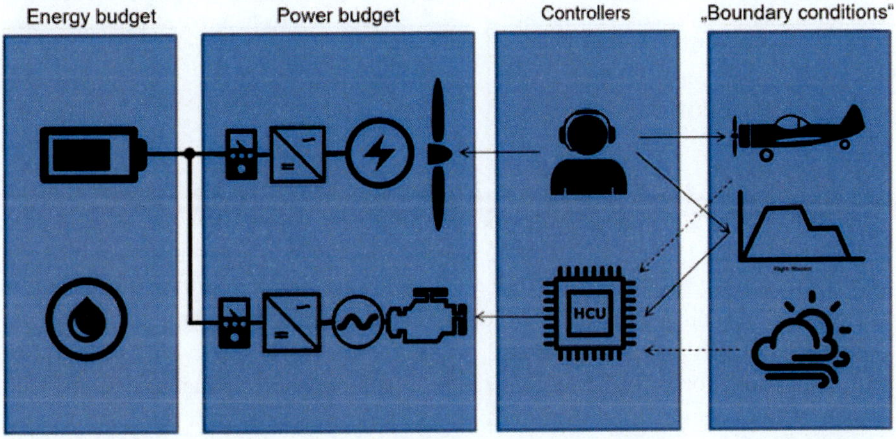

Fig. 4 Simplified overview of control schematic

3 Outlook

Flight testing of the described system began in 2021, but focused initially on the flight safety and general functionalities. Further studies and longer flights are needed to provide more results in terms of flight performance and efficiencies. In general, the predicted key specifications and significant reductions in carbon and noise emissions are easily achievable. All systems for operational safety are working as expected (Fig. 5).

Additionally, the use of a renewable fuel has been demonstrated in the aircraft with no resulting limitations on the system, but will need more test time to provide accurate results. Comparison to conventional fuels will be essential factor for the acceptance of sustainable fuels in aviation.

Fig. 5 e-Genius hybrid prepared for test flight campaign

4 Conclusion

A serial hybrid propulsion system has been implemented into a modern general aviation aircraft to demonstrate the possibility of a very low fuel consumption two-seater. The system has been developed and deemed airworthy by the German Federal Aviation Office. The implementation of this aircraft is presented and discussed. Findings apply to various flying platform including fixed-wing and other electrically propelled aircraft.

A combination of highly efficient aerodynamics enabled by an electric drivetrain with state-of-the-art hybrid propulsion technology results in a test platform with uniquely low fuel burn and low carbon emissions.

System automation provides safe and efficient operation of the hybrid propulsion system while not increasing pilot workload over the usual amount.

Renewable fuels for aviation purposes will most likely be more expensive compared to fossil fuels. A drastic reduction in fuel burn of typically more than 60% might be able to offset some or all of the cost increase of the sustainable alternatives and thus produce a net carbon neutral hybrid aircraft, which has none of the disadvantages of solely battery powered aircraft.

It has to be noted that the technology readiness level of those hybrid drivetrains is still too low for immediate certification and the implementation poses some serious design challenges to the developers, but ultimately it is demonstrated, that the described method is one viable option to produce more environmentally friendly aircraft today.

References

Bravo GM, Praliyev N, Veress Á (2021) Performance analysis of hybrid electric and distributed propulsion system applied on a light aircraft. Energy 214:15

Donateo T, et al (2017) Real world fuel consumption of a piston-prop aircraft. In: 7th EASN conference

Glassock R, Galea M, Williams W, Glesk T (2017) Hybrid electric aircraft propulsion case study for skydiving mission. Aerospace 4(3):45

ICAO (2019) Electric, hybrid, and hydrogen aircraft – state of Play. Environmental Reports 2019

Sziroczak D, Jankovics I, Gal I, Rohacs D (2020) Conceptual design of small aircraft with hybrid-electric propulsion systems. Energy 204:117937

Thermal Analysis of ASTINSAT-1

Alper Şanlı

1 Introduction

Cubesats are smaller in size and lighter in weight compared to conventional artificial satellites. They perform tasks similar to other satellites, but they can manage a smaller number of payloads (Aslan et al. 2014). With the increase in technology, many problems that exist are solved with the solutions provided from the space field, and this situation increases the use of cubesats. Increasing investment in this area has increased the variety of subsystems used in cubesats and facilitated access to subsystems. The number of launch vehicles suitable for cubesats is increasing and launch costs are decreasing. This paves the way for cubesats to reach space.

In this study, the thermal analysis of the Astin Sat 1 cubesat, which is planned to be built in the near future, has been made. Astin Sat 1 cubesat has 4 missions. Its first task is to obtain telemetry data with the help of sensors on the cubesat and transfer the telemetry data from the cubesat to the ground station. Its second task is to broadcast to the Earth from the space environment an information that was determined by the transmitter to be added to the cubesat subsystem and previously added to the cubesat. The third task is to read the orbital Magnetosphere information during the movement of the satellite with the sensor to be added to the cubesat subsystem, transfer it to the ground station, and create "Earth Magnetic Field Modeling" in the ground station. The last task is the task of acquiring an image from the ground surface at any time and transferring it to the ground station, thanks to the above-meter resolution camera to be added to the cubesat subsystem. Cubesat modeling has been made to fulfill these tasks. The final design of the cubesat, suitable for the tasks, was carried out by making improvements in the modeling with thermal analysis.

A. Şanlı (✉)
Hezârfen Aeronautics and Space Technologies Institute, National Defence University Istanbul, Istanbul, Turkey

T. H. Karakoc et al. (eds.), *Emerging Trends in Electric Aviation*, Sustainable Aviation, https://doi.org/10.1007/978-3-031-37299-5_3

2 Astin Sat 1 CubeSat

Cubesats generally consist of seven subsystems: structure subsystem, electrical power subsystem, communication subsystem, command and data management subsystem, attitude determination and control subsystem, thermal subsystem, and propulsion subsystem (propulsion system is not generally used) (Aslan et al. 2014). Subsystems are selected in accordance with the task. The determined subsystems were modeled and made suitable for analysis. Astin Sat 1 cubesat is in 3 U standards: width 100 mm, length 100 mm, and height 340.5 mm. The total mass of Astin Sat 1 is 3230 grams. Cubesat building material is aluminum 7075 T6; subsystems are generally FR-4 materials so that cards in PCB PC/104 standard form factor can be placed (Fig. 1).

Astin Sat 1 cubesat has an altitude of 400 km and an eccentricity of 0°. The probe rocket will be used during the actual launch. The orbital type may change with the change of rocket characteristics. Its orbit has inclination of 45°, ascending node longitude, true anomaly value, and perigee angle of 0°. These orbital parameters will deviate from their actual values after the cube at is launched into orbit.

2.1 Thermal Analysis Results of Astin Sat 1

Due to temperature differences in the space environment, heat transfer to the satellite structure takes place. Heat transfer can occur through conduction, convection, and radiation (Shinde et al. 2017). Total heat transfer Q is expressed as heat transfer by conduction $Q_{Conduction}$, heat transfer by convection $Q_{Convection}$, and heat transfer by radiation $Q_{Radiation}$. Conductive and radiative heat transfer are the main modes of transfer mechanisms within a spacecraft in space.

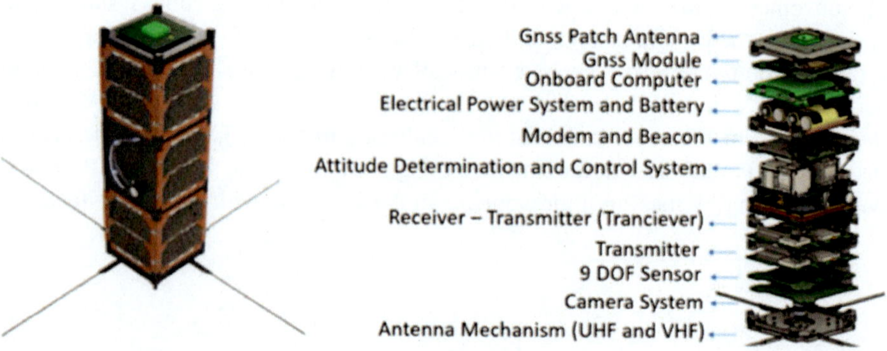

Fig. 1 Astin Sat 1 CubeSat example that can be placed on the launch vehicle (Şanlı & Aslan 2021)

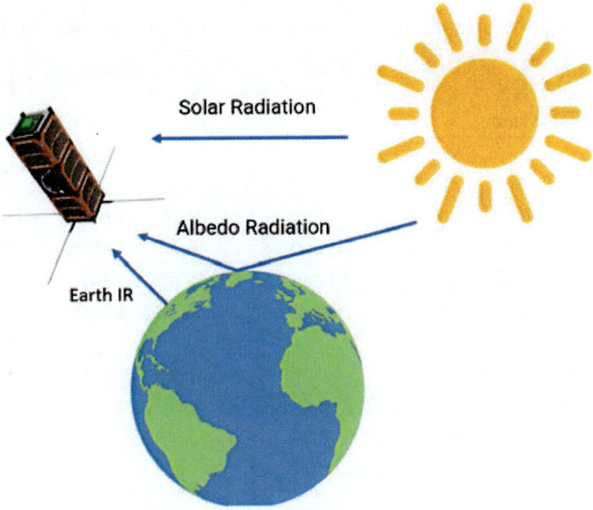

Fig. 2 Environmental factors encountered by the cubesat during its mission (Wertz et al. 2011)

$$Q = Q_{\text{Conduction}} + Q_{\text{Convection}} + Q_{\text{Radiation}} \tag{1}$$

Conduction is heat transfer due to collisions between particles. In Eq. 2, it is the conduction equation, and K is thermal conductivity in W/mK and A is cross-sectional area of the surface in m^2 (Garzón & Villanueva 2018).

$$\dot{Q}_{\text{Conduction}} = -KA\frac{(\Delta T)}{\Delta x} \tag{2}$$

Radiation is the energy emitted by matter and its conversion from thermal energy to electromagnetic energy. In Eq. 3, it is the radiation equation and ϵ is the emissivity of the surface; $\sigma = 5.67 \times 10^{-8}$ W/m^2 K^4 is the Stefan Boltzmann constant, A is the area of the surface in m^2, and T is the temperature in K (Fig. 2) (Brown 2002)

$$\dot{Q}_{\text{Radiation}} = \in \sigma A T^4 \tag{3}$$

Temperature analyses of the orbiting cubesat were carried out. Operating subsystem temperatures during the mission were analyzed. The suitability of the subsystems under the influence of these temperatures was investigated. Anodizing was applied to the materials. The W values and properties produced by the subsystems are as shown in Table 1.

In order to increase the durability of the cubesat against temperature, the materials are coated with anodized. The endurance temperatures of the subsystems during and outside the mission are given. In order for the Astin Sat 1 cubesat subsystems to work efficiently, they should not exceed these temperatures during the mission. Subsystems work at certain time intervals. While performing the thermal analysis,

Table 1 Thermal analysis temperature condition, material, and power consumption amount

Subsystems	Temperature during mission [°C]	Temperature during non-mission [°C]	Material	Power consumption [W]
Structure	−150 and +150	−150 and +150	Al 7075 T6 and anodized	0
Gnss patch antenna	−40 and +85	−50 and +90	FR4 and copper	0.05
Gnss module	−40 and +85	−50 and +90	FR4 and copper	1
Onboard computer	−40 and +85	−55 and +95	FR4 and copper	0.38
Electrical power system and battery	−20 and +60	−20 and +60	FR4 and copper	0.1
Modem and beacon	−30 and +80	−30 and +80	FR4 and copper	0.9
Attitude determination and control system	−20 and +70	−20 and +70	FR4 and copper	0.15–2.5
Receiver–transmitter (tranciever)	−25 and +60	−25 and +60	FR4 and copper	0.15–1.5
Transmitter	−25 and +60	−25 and +60	FR4 and copper	0.15–1.5
9 DOF sensor	−30 and +50	−30 and +50	FR4 and copper	1
Camera system	−30 and +50	−30 and +50	FR4 and copper	1
Antenna mechanism (UHF and VHF)	−100 and +120	−150 and +120	Al 7075 T6 and anodized	1.5
Solar panels	−40 and +100	−60 and +125	FR4	0.05

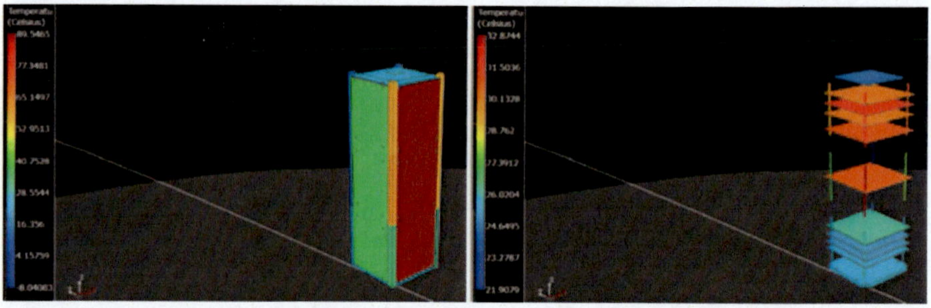

Fig. 3 The highest temperatures that Astin Sat 1 exterior and internal can encounter

the time intervals in which the subsystems work were taken into account. The highest and lowest temperature values that the cubesat can encounter in orbit has been calculated (Figs. 3 and 4).

The cubesat rose to a maximum of 89.5 °C in orbit. The solar panel, from which sunlight comes, is under the highest temperature effect. Satellite subsystems, on the other hand, reached a maximum temperature of 31 °C. The temperature at which the cubesat is coldest is −42.9 °C. The solar panel is under the influence of this temperature. The lowest temperature of satellite subsystems is −20 °C (Figs. 5 and 6).

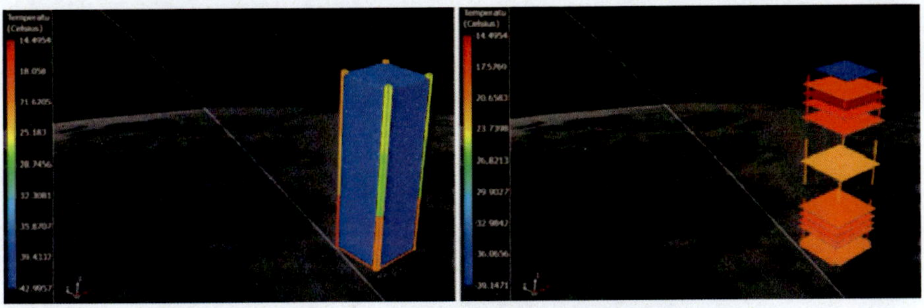

Fig. 4 Lowest temperatures that Astin Sat 1 exterior and internal can encounter

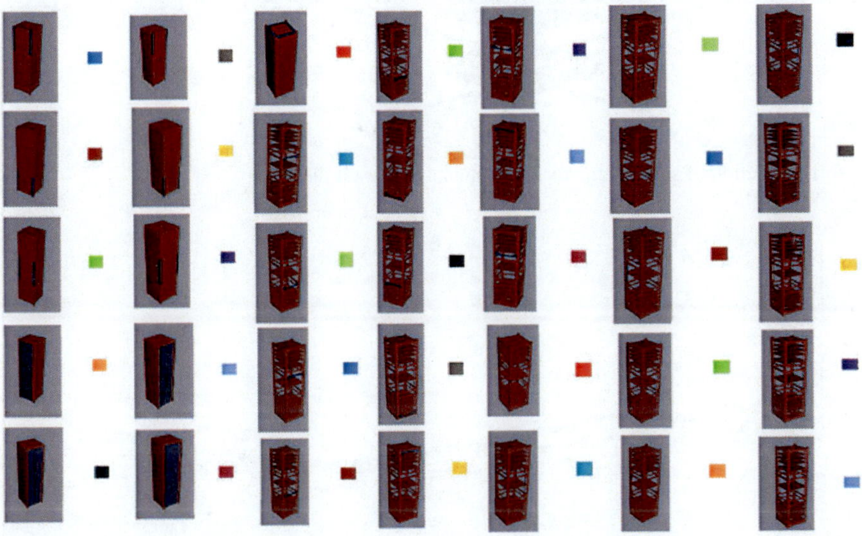

Fig. 5 Color distribution of hardware for thermal analysis result display

A full orbital orbit of the Astin Sat 1 cubesat was analyzed. Time-dependent temperature graph, time-dependent Earth albedo absorbed flux graph, and time-dependent solar absorbed flux graphs were obtained (Figs. 7, 8 and 9, Table 2).

3 Conclusion

The operating times of the subsystems were taken into account while performing the thermal analysis. As a result of the thermal analysis, the cubesat reached a maximum of 89.5 °C in orbit. The solar panel, from which sunlight comes, is under the highest temperature effect. Satellite subsystems other than the structure, on the other hand,

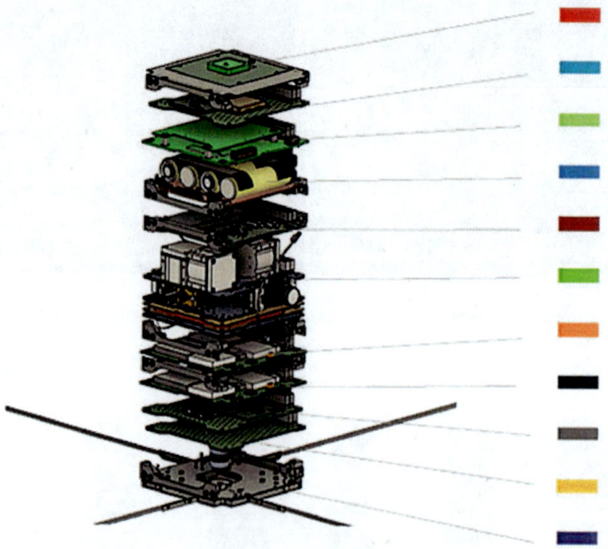

Fig. 6 Distribution of subsystems for thermal analysis result display

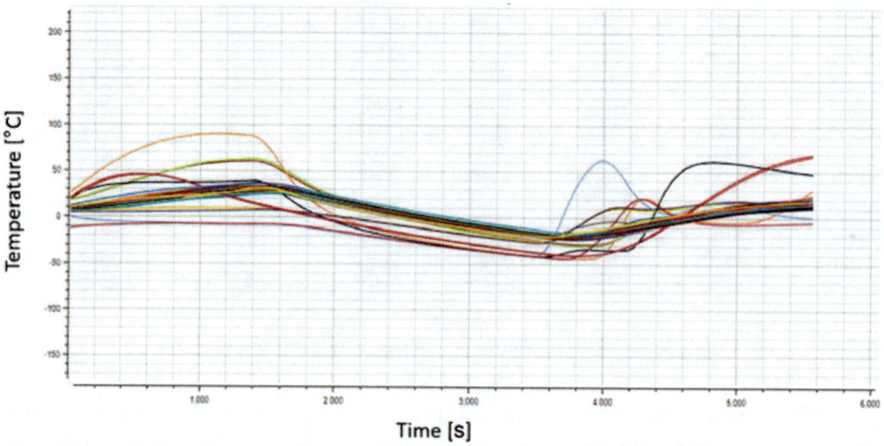

Fig. 7 Time-dependent temperature graph

reached a maximum temperature of 31 °C. The temperature at which the cubesat is coldest is −42.9 °C. The solar panel is under the influence of this temperature. The lowest temperature of the satellite subsystems other than the structure is −20 °C. As a result of the thermal analysis, the structure designed for the Astin Sat 1 cubesat and the selected subsystems are suitable for the missions in the specified orbit.

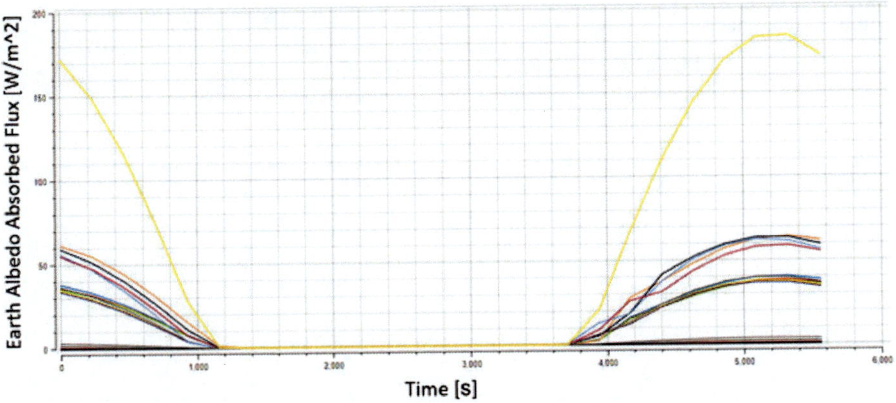

Fig. 8 Distribution of earth albedo absorb flux graph over time

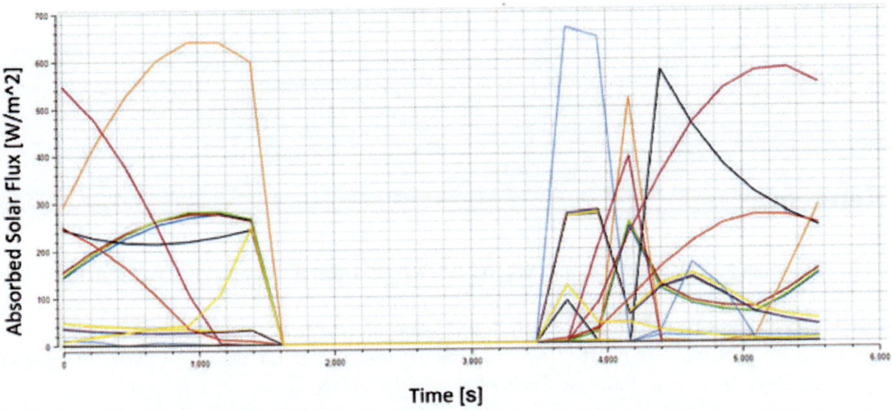

Fig. 9 Dependent absorption solar flux graph

Table 2 Thermal analysis temperature status and availability result

Subsystems	Minimum required temperature [°C]	Required highest temperature [°C]	Lowest measured temperature [°C]	Highest temperature measured [°C]	Result
Structure	−150	+150	−35	+75	Positive
Gnss patch antenna	−50	+90	−40	+30	Positive
Gnss module	−50	+90	−17	+30	Positive
Onboard computer	−55	+95	−14	+31	Positive
Electrical power system and battery	−30	+80	−18	+30	Positive

(continued)

Table 2 (continued)

Subsystems	Minimum required temperature [°C]	Required highest temperature [°C]	Lowest measured temperature [°C]	Highest temperature measured [°C]	Result
Modem and beacon	−20	+60	−18	+30	Positive
Attitude determination and control system	−20	+70	−20	+30	Positive
Receiver–transmitter (Tranciever)	−25	+60	−18	+26	Positive
Transmitter	−25	+60	−18	+26	Positive
9 DOF sensor	−30	+50	−19	+25	Positive
Camera system	−30	+50	−18	+25	Positive
Antenna mechanism (UHF and VHF)	−100	+120	−18	+25	Positive
Solar panels	−60	+125	−42.9	+89.5	Positive

References

Aslan AR, Ümit E, Şimşek M, Uludağ MŞ, Aksulu MD, Sofyalı A, Kalemci E (2014) Qb50 Projesi Kapsamında Beeaglesat Küp Uydusunun Geliştirilmesi

Brown CD (2002) Elements of spacecraft design. Aiaa

Garzón A, Villanueva YA (2018) Thermal analysis of satellite Libertad 2: a guide to cubesat temperature prediction. J Aerosp Technol Manag 10:e4918

Şanlı A, Aslan AR (2021) Design of ASTINSAT-1 and structural analysis, use of generative design. In: International congress on engineering and technology management

Shinde P, Quintero A, Tansel I, Tosunoglu S (2017) CubeSat thermal analysis. In: 30th Florida conference on recent advances in robotics, Florida Atlantic University, Boca Raton

Wertz JR, Everett DF, Puschell JJ (2011) Space mission engineering: the new SMAD. Microcosm Press

Numerical Examination of Different Flow Channel Fractions Effects in a Vanadium Redox Flow Battery with Serpentine Flow Field

Ilker Kayali

1 Introduction

Among the energy storage technologies, vanadium redox flow batteries have attracted a lot of attention in terms of scalable storage capacity, high energy density (Tsushima and Suzuki 2020), and use with renewable energy sources (Jiao et al. 2022). The VRFB electrode potential may vary according to geometric structure (Yin et al. 2014). The importance of the ratio of channel height to cell width for the VRFB system is understood (Tsushima and Suzuki 2020). In a VRFB system, large changes occurred in electrode and electrolyte potential from the inlet channel to the outlet channel (Cheng et al. 2020).

This study aims to examine the effect of charge and mass transports in the serpentine flow field for four different channel fractions in the discharge process. In numerical models, while the channel heights remain at a constant value, each channel height is changed.

This study examines the effects of changes in the channel fractions, electrode/electrolyte potentials, and velocity/pressure distributions. To perform this work, the two-dimensional steady state is obtained using COMSOL Multiphysics 5.5.

I. Kayali (✉)
Department of Energy Systems Engineering, Erciyes University, Kayseri, Turkey

Kapadokya Vocational School, Kapadokya University, Nevşehir, Turkey
e-mail: ilker.kayali@kapadokya.edu.tr

T. H. Karakoc et al. (eds.), *Emerging Trends in Electric Aviation*, Sustainable Aviation, https://doi.org/10.1007/978-3-031-37299-5_4

2 Method

The VRFB schematic diagram consisting of an electrode, membrane, current collector layer, and electrolyte tanks is shown in Fig. 1. The geometric properties of the two-dimensional validated model are given in Table 1. The model equations are solved under the specific boundary conditions according to the following assumptions:

- The phenomenon is steady-state.
- The flow is incompressible.
- The cell is isothermal.
- Only H^+ ions are allowed to pass through the membrane.
- The are no side reactions.
- The porous electrodes and membrane are assumed isotropic and homogeneous.
- The contact resistances at interfaces are ignored.
- The evolutions of hydrogen and oxygen in the membrane are neglected.

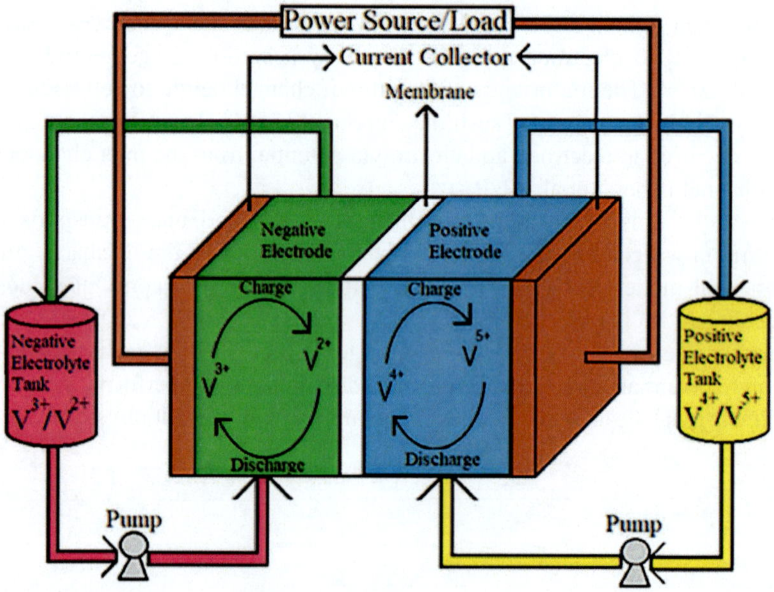

Fig. 1 A Schematic diagram of a single cell VRFB

Table 1 The geometric properties of the components used for the model

Symbols	Definition	Value
L_e	Electrode thickness	6 (mm)
L_m	Membrane thickness	183 (μm)
W_{ch}	Channel height	4 (mm)
H_{ch}	Channel thickness	1.59 (mm)
W_{rib}	Rib height	7.50 (mm)
W_{cell}	Electrode width	50 (mm)

2.1 Model Equations

The model is made up of continuity and conservation equations (mass, charge, momentum, and species) (Al-Yasiri and Park 2017). The channel fractions obtained by only the channel height change are given in Eq. 1. The channel fractions for the serpentine flow field are shown in Fig. 2.

$$\lambda = \frac{W_{\text{ch}}}{W_{\text{rib}} + W_{\text{ch}}} \tag{1}$$

The channel fractions and numerical parameter values are shown in Table 2. The parameter values for the numerical model are taken from the literature (Al-Yasiri and Park 2017).

2.2 Boundary Conditions

The boundary conditions are applied to the model equations containing the continuity equation, conservation of momentum, mass, species, and charges (Al-Yasiri and Park 2017).

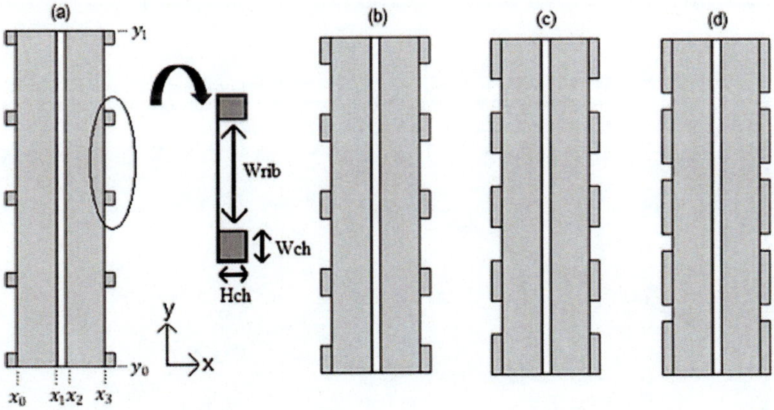

Fig. 2 The boundary conditions in channel fractions for serpentine flow field; **(a)** λ_1=0.2, **(b)** λ_2=0.4, **(c)** λ_3=0.6, **(d)** λ_4=0.8

Table 2 Channel fractions

λ	W_{ch}	$W_{rib} + W_{ch}$	$W_{ch}/(W_{rib}+W_{ch})$
λ_1	2 × 5 (mm)	5 × 10^{-2} (m)	0.20
λ_2	4 × 5 (mm)	5 × 10^{-2} (m)	0.40
λ_3	6 × 5 (mm)	5 × 10^{-2} (m)	0.60
λ_4	8 × 5 (mm)	5 × 10^{-2} (m)	0.80

3 Results and Discussions

The numerical solution for the serpentine flow field is validated against the experimental data in the literature (Al-Yasiri and Park 2017), as shown in Fig. 3. It is observed that the average relative error between the experimental data and the numerical model is less than 1% during discharge. The charge result is well-fitted with an average relative error of less than 3%.

The contour results obtained are for the discharge process at 0.5 SOC. The effect of the channel fractions on the electrode potential is given in Figs. 4a and 4b. The electrode potential is increased throughout the membrane for both the negative and positive electrodes.

In addition, the electrode potential occurred in channel fractions of 0.8, 0.6, 0.4, and 0.2, respectively, from high to low.

The effect of different channel fractions on the electrolyte potential is shown in Figs. 5a and 5b. As the channel fraction rises, the electrolyte potential at the electrode increases.

The effects of channel fractions on velocity and pressure distributions are given in Figs. 6a and 6b. The highest velocity values occurred in the 0.8 channel fraction, while the lowest velocity values occurred in the 0.2 channel fraction. Moreover, the pressure value reaches 7200 Pa in the 0.2 channel fraction, while it reaches 4700 Pa in the 0.8 channel fraction.

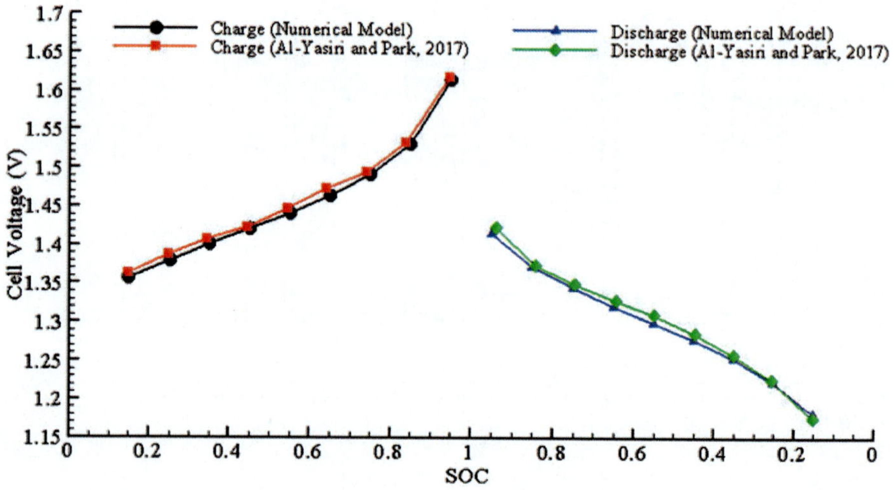

Fig. 3 Comparison of experimental results and simulation data

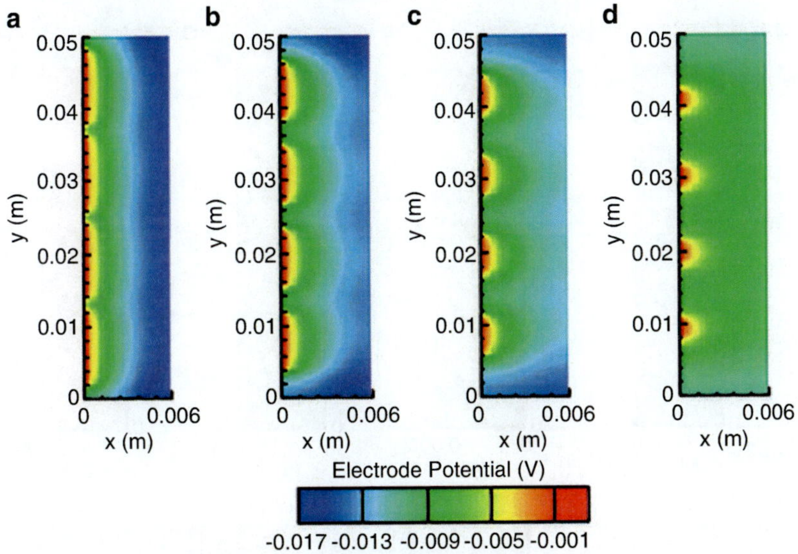

Fig. 4a Effect of different channel fractions negative electrode potential for discharge process; (**a**) λ_1=0.2, (**b**) λ_2=0.4, (**c**) λ_3=0.6, (**d**) λ_4=0.8

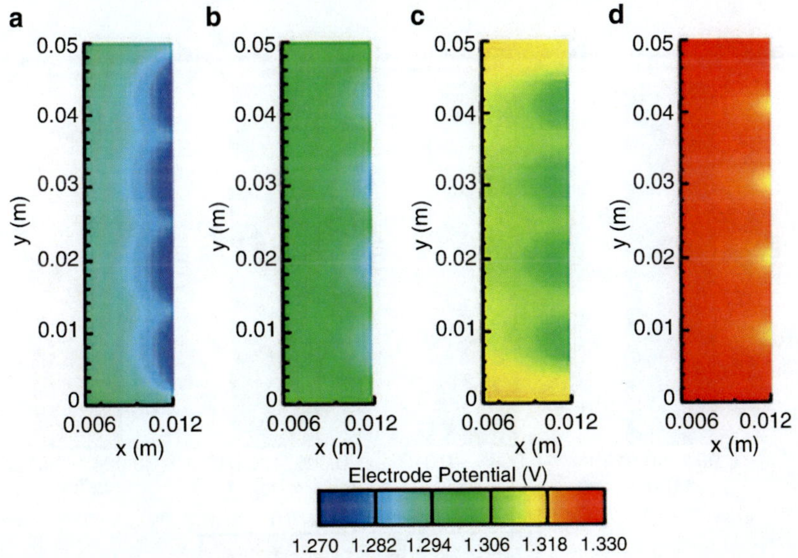

Fig. 4b Effect of different channel fractions positive electrode potential for discharge process; (**a**) λ_1=0.2, (**b**) λ_2=0.4, (**c**) λ_3=0.6, (**d**) λ_4=0.8

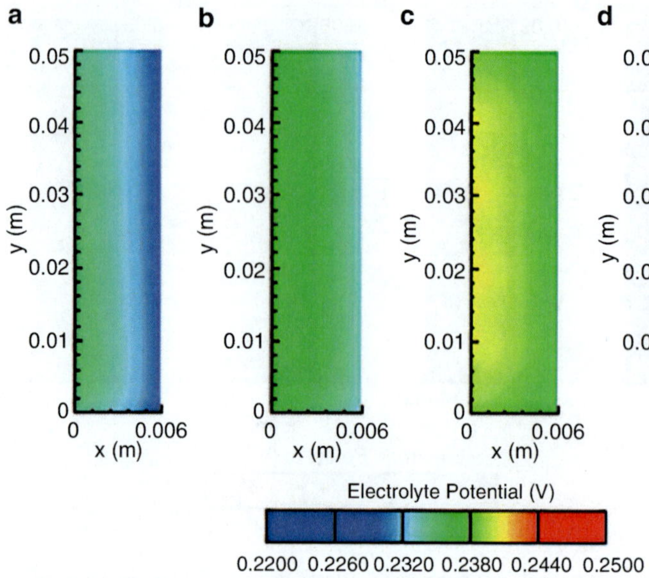

Fig. 5a Effect of different channel fractions negative electrolyte potential for discharge process; (**a**) $\lambda_1=0.2$, (**b**) $\lambda_2=0.4$, (**c**) $\lambda_3=0.6$, (**d**) $\lambda_4=0.8$

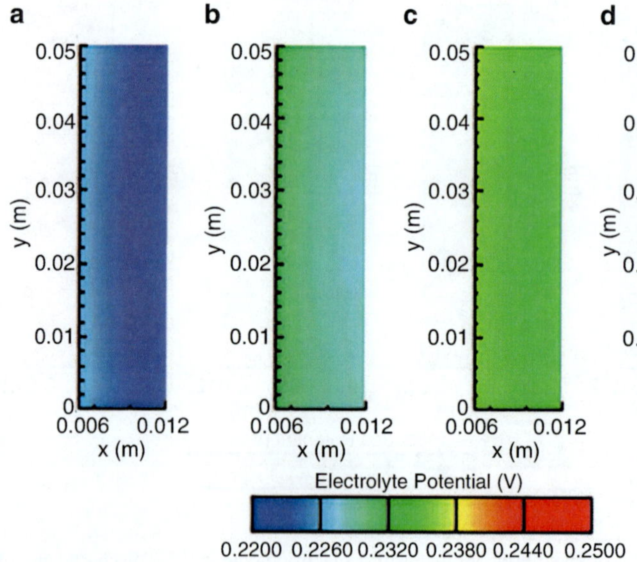

Fig. 5b Effect of different channel fractions positive electrolyte potential for discharge process; (**a**) $\lambda_1=0.2$, (**b**) $\lambda_2=0.4$, (**c**) $\lambda_3=0.6$, (**d**) $\lambda_4=0.8$

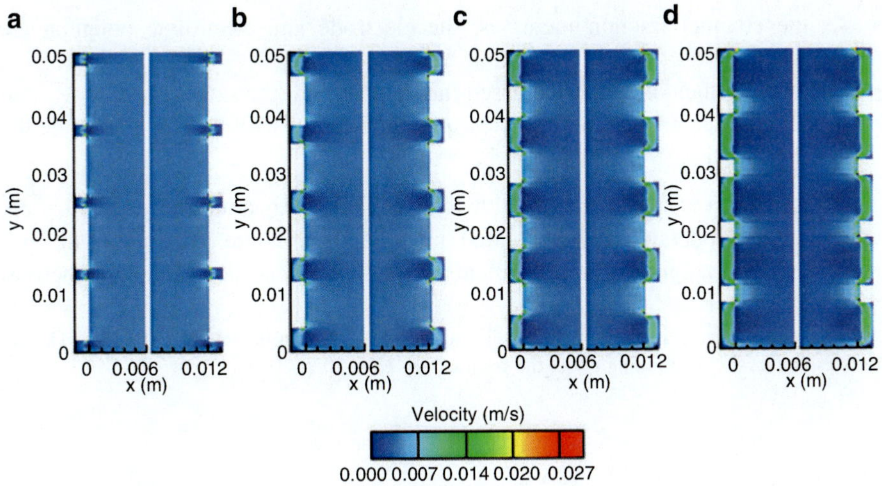

Fig. 6a Effect of different channel fractions velocity distribution for discharge process; (**a**) $\lambda_1=0.2$, (**b**) $\lambda_2=0.4$, (**c**) $\lambda_3=0.6$, (**d**) $\lambda_4=0.8$

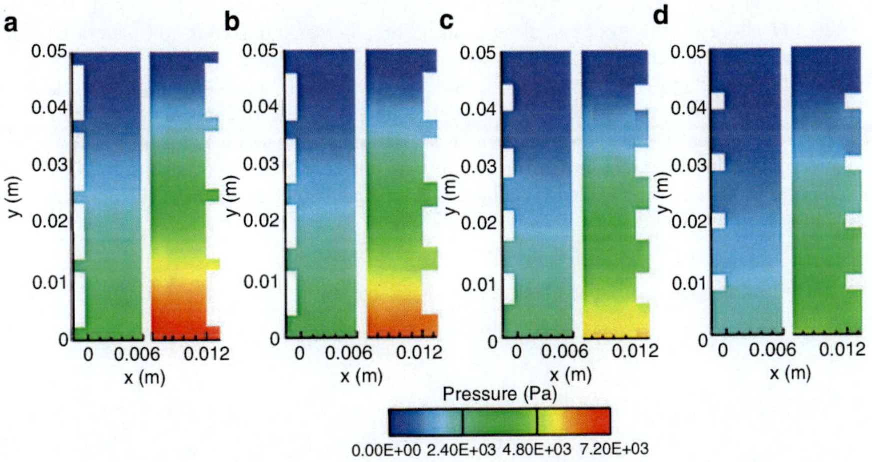

Fig. 6b Effect of different channel fractions pressure distribution for discharge process; (**a**) $\lambda_1=0.2$, (**b**) $\lambda_2=0.4$, (**c**) $\lambda_3=0.6$, (**d**) $\lambda_4=0.8$

4 Conclusion

The aim of this study is to analyze the effects of the electrode/electrolyte potential and velocity/pressure distribution on different channel fractions in both negative and positive electrodes in a vanadium redox flow battery. The channel fractions of 0.2, 0.4, 0.6, and 0.8 are created for the serpentine flow field. The results from this study are as follows:

- As the channel fraction increases, the electrode and electrolyte potential are increased.
- The rise in channel height increased the velocity values.
- In the 0.2 channel fraction, the pressure value reaches 7200 Pa, while in the 0.8 channel fraction, it reaches 4700 Pa.

As a result, it is shown that the flow channels of a vanadium redox flow battery affect the electrode/electrolyte potential and velocity/pressure distribution. In summary, it will be helpful for improvements in the flow fields and channel geometries.

Acknowledgments The authors would like to thank the Scientific Research Projects Unit of Erciyes University for funding under the contract no: FYL-2020-10397.

References

Al-Yasiri M, Park J (2017) Study on channel geometry of all-vanadium redox flow batteries. J Electrochem Soc 164:A1970

Cheng Z, Tenny KM, Pizzolato A, Forner-Cuenca A, Verda V, Chiang YM, Brushett Y, Behrou R (2020) Data-driven electrode parameter identification for vanadium redox flow batteries through experimental and numerical methods. Appl Energy 279:115530

Jiao YH, Lu MY, Yang WW, Tang XY, Ye M, Xu Q (2022) A 3D macro-segment network model for vanadium redox flow battery with serpentine flow field. Electrochim Acta 403:139657

Tsushima S, Suzuki T (2020) Modeling and simulation of vanadium redox flow battery with interdigitated flow field for optimizing electrode architecture. J Electrochem Soc 167(2):020553

Yin C, Gao Y, Guo S, Tang H (2014) A coupled three dimensional model of vanadium redox flow battery for flow field designs. Energy 74:886–895

Flutter Analysis of a 3-D Box Wing with Distributed Electric Propulsion

S. Ahmad Fazelzadeh, Abbas Mazidi, and Amirhossein Ghasemikaram

Nomenclature

a	Non-dimensional parameter related to elastic axis of airfoil
E	Elastic module
G	Torsional module
H	Heaviside function
I	Moment of inertia
J	Polar moment of inertia
k_l	Stiffness of longitudinal spring
k_t	Stiffness of torsional spring
l	Length of the wing
L	Lift force
m	Mass per unit length
M	Aerodynamic moment
M_e	Mass of motor
p	Follower force
T_t	Total kinetic energy
w	Bending deflection
$W_{n.c}$	Non-conservative done work
x_θ	Distance between center of gravity and elastic axis
δ_D	Dirac Delta function
Λ	Sweep angle

S. A. Fazelzadeh
Shiraz University, Shiraz, Iran
e-mail: Fazelzad@shirazu.ac.ir

A. Mazidi (✉) · A. Ghasemikaram
Yazd University, Yazd, Iran
e-mail: Amazidi@yazd.ac.ir; Ah.ghasemikaram@stu.yazd.ac.ir

T. H. Karakoc et al. (eds.), *Emerging Trends in Electric Aviation*, Sustainable Aviation, https://doi.org/10.1007/978-3-031-37299-5_5

θ	Torsional deflection
f	Front index
r	Rear index

1 Introduction

Employing distributed electric propulsion, which involves distributing electric motors and associated propulsions along the aircraft wing, can improve performance in noise reduction, aerodynamic efficiency, and flight safety. Although electric propulsion for aircraft applications forms a very small portion of current standard aviation due to its limitations, but this technology has the potential to be integrated into a wide number of future aircrafts. Mounting all motors to the box wing is a natural consideration, but because of the propellers' thrust and electric motors mass and the wing deformations, it might be subjected to the aeroelastic problems. So, aeroelastic analysis of DEP aircraft wings could be very challenging and important task for design of future electrical aircraft.

The bending-torsional flutter of an un-swept wing was investigated by Feldt and Herrmann (1974). The wing was modeled as a cantilever beam subjected to a follower force which represented the engine thrust. The flutter and divergence phenomena of beams and plates are subjected to non-conservative forces studied by Zuo and Schreyer (1996). The results exhibit that the plate instability is governed by flutter for a specified range of the non-conservative parameters. The effects of bending to torsional rigidities ratio are evaluated on the stability boundary of a straight clean wing subjected to a transverse follower force (Hodges et al. 2002). A parametric study on the effects of follower force and external store on the flutter speed and frequency of an un-swept wing are investigated as Ref. (Fazelzadeh et al. 2009). The numerical results showed that the store location and magnitude and thrust force affect the wing flutter boundary. The aeroelastic stability of a swept wing carrying a powered engine is assessed by Fazelzadeh and Mazidi (2010); Mazidi and Fazelzadeh (2013). In order to become possible in view of "green aviation," a distributed propulsion system is recommend by Frediani et al. (2012). The bending-torsional flutter of a composite cantilever wing subjected to a thrust force is presented by Amoozgar and Irani (2012). The flutter boundary of swept wings carrying twin powered engines is predicted by Mazidi and Fazelzadeh (2013). The numerical simulation showed that some design parameters such as sweep angle, mass ratio, and engine locations can influence on the flutter velocity and frequency. The effects of location, mass, and inertia of a store on the aeroelastic time response of an aircraft wing are demonstrated as Ref. (Xu and Ma 2014). The analytical nonlinear flutter and sensitivity analysis of high aspect ratio wings subjected to a follower force are assessed by Zafari et al. (2018). The effects of distributed follower force on the aeroelastic stability of wing are studied by Amoozgar et al. (2021). The electric motors had different properties and two case studies are considered.

Numerical results demonstrate that the tip follower force, mass, and angular momentum had the most impact on the aeroelastic stability. In order to postpone the flutter of a high aspect ratio wing, distributed electric propulsion is considered by Bamberger (2021). The use of distributed engines to improve aeroelastic behavior is tried by Memmolo et al. (2022). Numerical results detect that the engine placement strongly enhance the flutter velocity and how the tip propeller can reduce the flutter speed.

Therefore, there is not yet enough literature studying the effects of distributed propulsion on the flutter of BWA configurations. The results of this paper can determine the effect of design parameters on the flutter stability of the box wings.

2 Governing Equations

In the present work, the BWA configuration includes a front wing with positive sweep and dihedral angles and a rear wing with negative sweep angle, and also the connection winglet is simulated by two longitudinal and torsional springs. Four electric motors are fixed on each wing at dimensionless intervals 0.2, 0.4, 0.6, and 0.8 as Fig. 1.

In order to derive the governing equations, several intermediate coordinate systems are utilized as Ref. (Ghasemikaram et al. 2022). The resulting equations for a clean BWA are extracted through Hamilton's variational principle and presented in Ref. (Ghasemikaram et al. 2022). On the other hand, due to presence of the motors on the wings, some terms are added to the kinetic energy and virtual work and are expressed as follows:

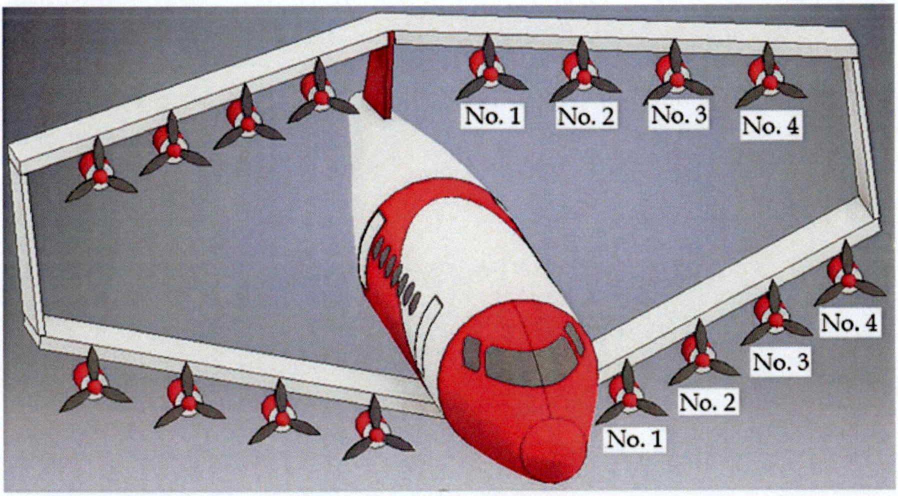

Fig. 1 A schematic of BWA with four electric motors on each wing

$$\delta T_t = \begin{array}{l} \frac{1}{2}\sum_{i=1}^{4}\left\{\int_0^{l_f}\iint_{A_{ef}} m_{e_{f_i}}\dot{R}_{e_{f_i}}.\delta\dot{R}_{e_{f_i}}\delta_D\left(x_{f_i}-l_f\right)dA_{e_{f_i}}\,dx_f\right. \\ \left.+\frac{1}{2}\int_0^{l_r}\iint_{A_{er}} m_{e_{r_i}}\dot{R}_{e_{r_i}}.\delta\dot{R}_{e_{r_i}}\delta_D(x_{r_i}-l_r)dA_{e_{r_i}}\,dx_r\right\} \end{array} \tag{1}$$

$$\delta W_{nc} = \sum_{i=1}^{4}\left\{\int_0^{l_f}\left[p_{f_i}\cdot\delta R_{ef_i}\right]dx_f + \int_0^{l_r}\left[p_{r_i}\cdot\delta R_{er_i}\right]dx_r\right\} \tag{2}$$

which $\dot{R}_e$ and p are position vectors of the engines C.G. and follower forces, respectively. Therefore, the final equations are derived as follows:

$$\begin{aligned}
&m_f\ddot{w}_f + m_f x_{\theta f}\ddot{\theta}_f + (EI)_f w_f^{(4)} \\
&+\int_0^{l_r}\left\{k_l w_f\left[\delta_D\left(x_f-l_f\right)\right]^3\delta_D(x_r-l_r) - k_l a_{33} w_r\left[\delta_D\left(x_f-l_f\right)\right]^2[\delta_D(x_r-l_r)]^2\right\} \\
&dx_r + \sum_{i=1}^{N}\{[M_{e_{f_i}}\ddot{w}_f + M_{e_{f_i}}y_{e_{f_i}}\cos\Lambda_f\ddot{\theta}_f - I_{e_{f_i}}\sin^2\Lambda_f\ddot{w}_f'' \\
&- M_{e_{f_i}}y_{e_{f_i}}{}^2\sin^2\Lambda_f\ddot{w}_f'' - M_{e_{f_i}}z_{e_{f_i}}{}^2\ddot{w}_f'' - M_{e_{f_i}}y_{e_{f_i}}{}^2\sin\Lambda_f\cos\Lambda_f\ddot{\theta}_f' - I_{e_{f_i}} \\
&\sin\Lambda_f\cos\Lambda_f\ddot{\theta}_f'\}]\delta_D\left(x_f - x_{e_{f_i}}\right)\} \\
&+\sum_{i=1}^{N}\{-p_{f_i}\left(x_{e_{f_i}}-x_f\right)H\left(x_{e_{f_i}}-x_f\right)\cos\Lambda_f\theta_f'' + 2p_{f_i}H\left(x_{e_{f_i}}-x_f\right)\cos\Lambda_f\acute{\theta}_f \\
&+p_{f_i}H\left(x_{e_{f_i}}-x_f\right)\sin\Lambda_f w_f'' + p_{f_i}\sin\Lambda_f\delta_D\left(x_f-x_{e_{f_i}}\right)\acute{w}_f\} = L_f
\end{aligned} \tag{3}$$

$$\begin{aligned}
&I_f\ddot{\theta}_f + m_f x_{\theta f}\ddot{w}_f - (GJ)_f\theta_f'' + \int_0^{l_r}\left\{k_t\theta_f\left[\delta_D\left(x_f-l_f\right)\right]^3\delta_D(x_r-l_r)\right. \\
&-k_t a_{11}\theta_r\left[\delta_D\left(x_f-l_f\right)\right]^2[\delta_D(x_r-l_r)]^2\}dx_r + \sum_{i=1}^{N}\left\{\left[M_{e_{f_i}}y_{e_{f_i}}\cos\Lambda_f\ddot{w}_f\right.\right. \\
&+I_{e_{f_i}}\cos^2\Lambda_f\ddot{\theta}_f + M_{e_{f_i}}y_{e_{f_i}}{}^2\cos^2\Lambda_f\ddot{\theta}_f + M_{e_{f_i}}z_{e_{f_i}}{}^2\ddot{\theta}_f \\
&+M_{e_{f_i}}y_{e_{f_i}}{}^2\sin\Lambda_f\cos\Lambda_f\ddot{w}_f'' + I_{e_{f_i}}\sin\Lambda_f\cos\Lambda_f\ddot{w}_f'']\delta_D\left(x_f-x_{e_{f_i}}\right)\} \\
&+\sum_{i=1}^{N}\left\{-p_{f_i}\left(x_{e_{f_i}}-x_f\right)H\left(x_{e_{f_i}}-x_f\right)\cos\Lambda_f w_f''\right. \\
&+p_{f_i}z_{e_{f_i}}\cos\Lambda_f\delta_D\left(x_f-x_{e_{f_i}}\right)\} = M_f
\end{aligned} \tag{4}$$

$$
\begin{aligned}
& m_r\ddot{w}_r + m_r x_{\theta r}\ddot{\theta}_r + (EI)_r w_r^{(4)} + \int_0^{l_f} \Big\{ k_l {a_{31}}^2 w_r [\delta_D(x_r - l_r)]^3 \delta_D(x_f - l_f) \\
& + k_l {a_{32}}^2 w_r [\delta_D(x_r - l_r)]^3 \delta_D(x_f - l_f) + k_l {a_{33}}^2 w_r [\delta_D(x_r - l_r)]^3 \delta_D(x_f - l_f) \\
& - k_l a_{33} w_f [\delta_D(x_f - l_f)]^2 [\delta_D(x_r - l_r)]^2 \Big\} dx_f + \sum_{i=1}^{N} \Big\{ \Big[M_{e_{r_i}} \ddot{w}_r \\
& + M_{e_{r_i}} y_{e_{r_i}} \cos\Lambda_r \ddot{\theta}_r - I_{e_{r_i}} \sin^2\Lambda_r \ddot{w}_r'' - M_{e_{r_i}} {y_{e_{r_i}}}^2 \sin^2\Lambda_r \ddot{w}_r'' \\
& - M_{e_{r_i}} {z_{e_{r_i}}}^2 \ddot{w}_r'' - M_{e_{r_i}} {y_{e_{r_i}}}^2 \sin\Lambda_r \cos\Lambda_r \ddot{\theta}_r' \\
& - I_{e_{r_i}} \sin\Lambda_r \cos\Lambda_r \ddot{\theta}_r' \Big] \delta_D\left(x_r - x_{e_{r_i}}\right) \Big\} + \sum_{i=1}^{N} \Big\{ -p_{r_i} \left(x_{e_{r_i}} - x_r\right) H\left(x_{e_{r_i}} - x_r\right) \cos\Lambda_r \theta_r'' \\
& + 2 p_{r_i} H\left(x_{e_{r_i}} - x_r\right) \cos\Lambda_r \acute{\theta}_r + p_{r_i} H\left(x_{e_{r_i}} - x_r\right) \sin\Lambda_r w_r'' \\
& + p_{r_i} \sin\Lambda_r \delta_D\left(x_r - x_{e_{r_i}}\right) \acute{w}_r \Big\} = L_r
\end{aligned} \tag{5}
$$

$$
\begin{aligned}
& I_r\ddot{\theta}_r + m_r x_{\theta r}\ddot{w}_r - (GJ)_r \theta_r'' + \int_0^{l_f} \Big\{ - k_t a_{11} \theta_f [\delta_D(x_f - l_f)]^2 [\delta_D(x_r - l_r)]^2 \\
& - k_t ({a_{11}}^2 + {a_{12}}^2 + {a_{13}}^2) \theta_r [\delta_D(x_r - l_r)]^3 \delta_D(x_f - l_f) \Big\} dx_f \\
& + \sum_{i=1}^{N} \Big\{ \Big[M_{e_{r_i}} y_{e_{r_i}} \cos\Lambda_r \ddot{w}_r + I_{e_{r_i}} \cos^2\Lambda_r \ddot{\theta}_r + M_{e_{r_i}} {y_{e_{r_i}}}^2 \cos^2\Lambda_r \ddot{\theta}_r \\
& + M_{e_{r_i}} {z_{e_{r_i}}}^2 \ddot{\theta}_r + M_{e_{r_i}} {y_{e_{r_i}}}^2 \sin\Lambda_r \cos\Lambda_r \ddot{w}_r'' \\
& + I_{e_{r_i}} \sin\Lambda_r \cos\Lambda_r \ddot{w}_r'' \Big] \delta_D\left(x_r - x_{e_{r_i}}\right) \Big\} \\
& + \sum_{i=1}^{N} \Big\{ -p_{r_i} \left(x_{e_{r_i}} - x_r\right) H\left(x_{e_{r_i}} - x_r\right) \cos\Lambda_r w_r'' + p_{r_i} z_{e_{r_i}} \cos\Lambda_r \delta_D\left(x_r - x_{e_{r_i}}\right) \Big\} \\
& = M_r
\end{aligned} \tag{6}
$$

3 Numerical Results

Schiktanz (2011) covered the conceptual design of a BWA configuration that able to transfer 250 passengers (Pr250) and considered the structure of the front wing according to the airbus 320 wing. The geometrical and physical characteristics of the BWA are mentioned as Refs. (Schiktanz 2011 and Ghasemikaram et al. 2022), and also the following non-dimensional parameters are made in order to solve the equations:

$$V = \frac{U}{b_f \omega_{\theta f}} \text{ and } P = p\frac{l^2}{\sqrt{GJ.EI}} \tag{7}$$

which U, GJ, EI, and $\omega_{\theta f}$ parameters are air velocity, torsional stiffness, bending stiffness, and the first torsional mode frequency of the front wing, respectively. Due to the existence of few time-dependent and parameter-dependent integral terms in the final PIDEs, a novel methodology is applied to the equations which includes several steps and is described in Ref. (Ghasemikaram et al. 2022). The arrangement of engines shutdown is considered in the five situations. It should be noted that the mass of each motor is considered $\frac{1}{8}$ of the wing mass.

- First situation: All eight motors are working.
- Second situation: Three motors are working on each wing (six motors in total).
- Third situation: Four motors are working on one wing (four motors in total).
- Forth situation: Two motors are working on each wing (four motors in total).
- Fifth situation: One motor is working on each wing (a total of two motors).

The motor numbering starts the root of the wings as present in Fig. 1. The flutter and divergence boundaries are demonstrated for the first situation in Fig. 2. As expected, the divergence boundary is greater than the flutter boundary, remarkably.

The effects of follower force on the flutter speed are observed for the second situation in Fig. 3. The stability boundary is maximum when the motors No. 1-2-3 are working, and the minimum flutter speed occurs when the motors No. 2-3-4 are working.

In the third situation (when the motors No. 1-2-3-4 are working in rear or front wing), for low values of electric engines thrust, the greatest stability boundary occurs when the front wing motors are working, and also for the higher values of electric

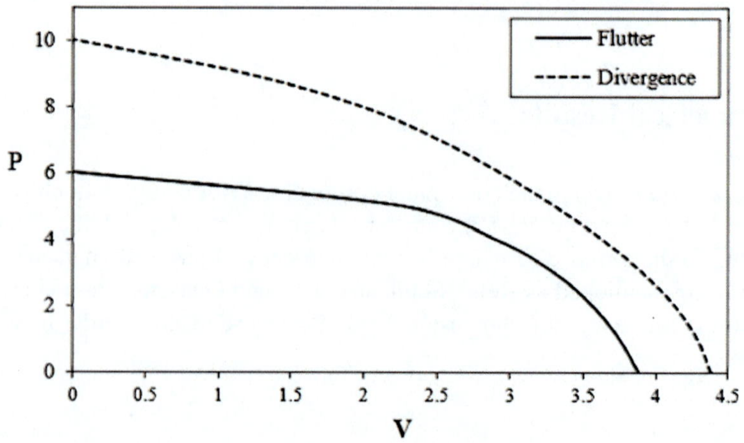

Fig. 2 The stability boundary for the first situation ($X_{e1} = 0.2, X_{e2} = 0.4, X_{e3} = 0.6, X_{e4} = 0.8, Y_{e(1,2,3,4)} = Z_{e(1,2,3,4)} = 0$)

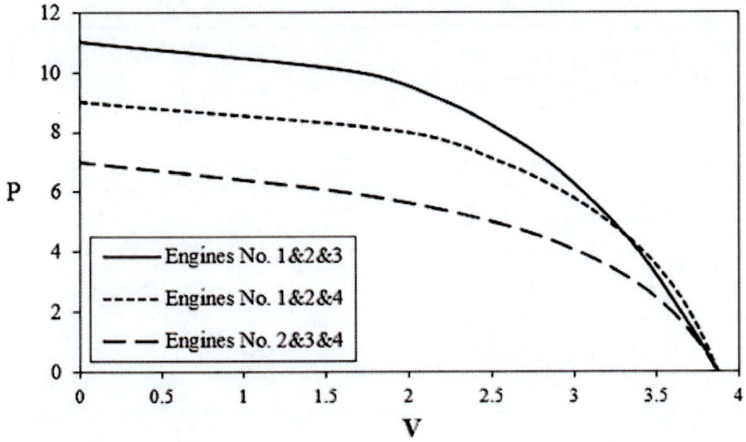

Fig. 3 The stability boundary for the second situation ($X_{e1} = 0.2, X_{e2} = 0.4, X_{e3} = 0.6, X_{e4} = 0.8$, $Y_{e(1,2,3,4)} = Z_{e(1,2,3,4)} = 0$)

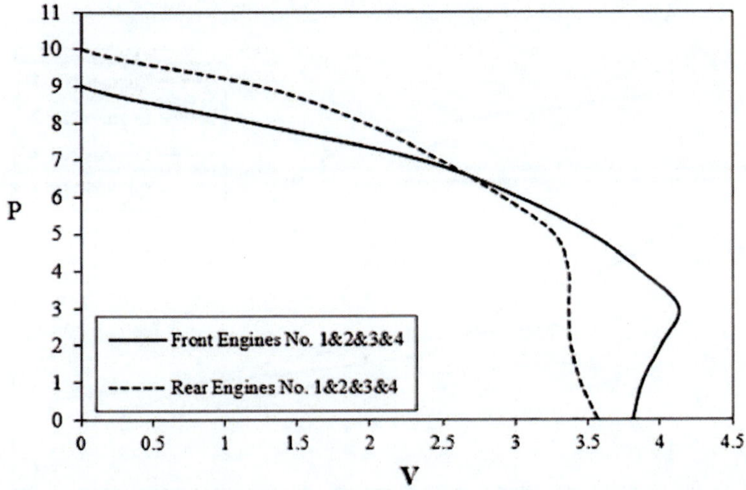

Fig. 4 The stability boundary for the third situation ($X_{e1} = 0.2, X_{e2} = 0.4, X_{e3} = 0.6, X_{e4} = 0.8$, $Y_{e(1,2,3,4)} = Z_{e(1,2,3,4)} = 0$)

engines thrust, the greatest stability boundary occurs when the rear wing motors are working, as shown in Fig. 4.

Figure 5 shows the effects of engine working arrangements for the fourth situation. The maximum and minimum stability regions are observed for the cases where engines No. 1-3 and 3–4 are working, respectively. The results show that engine No. 1 has the most important role in this situation.

The stability boundaries are shown for the fifth situation in Fig. 6. The maximum stability region is seen in the case where motor No. 2 is working, and on the other hand, the minimum stability region is observed in the case where motor No. 4 is working.

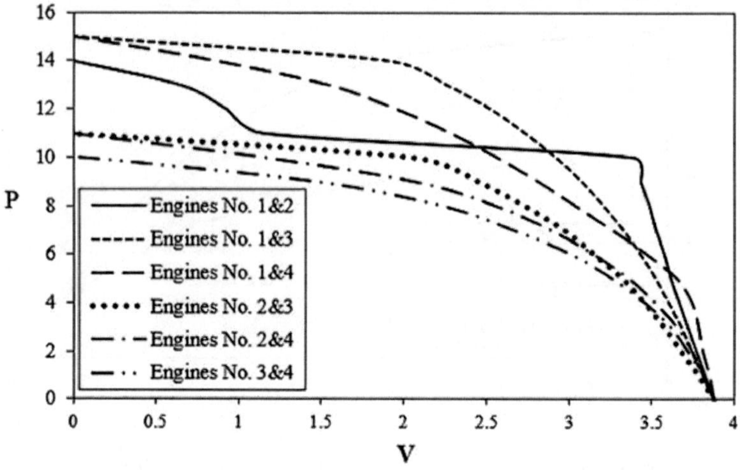

Fig. 5 The stability boundary for the forth situation ($X_{e1} = 0.2, X_{e2} = 0.4, X_{e3} = 0.6, X_{e4} = 0.8, Y_{e(1,2,3,4)} = Z_{e(1,2,3,4)} = 0$)

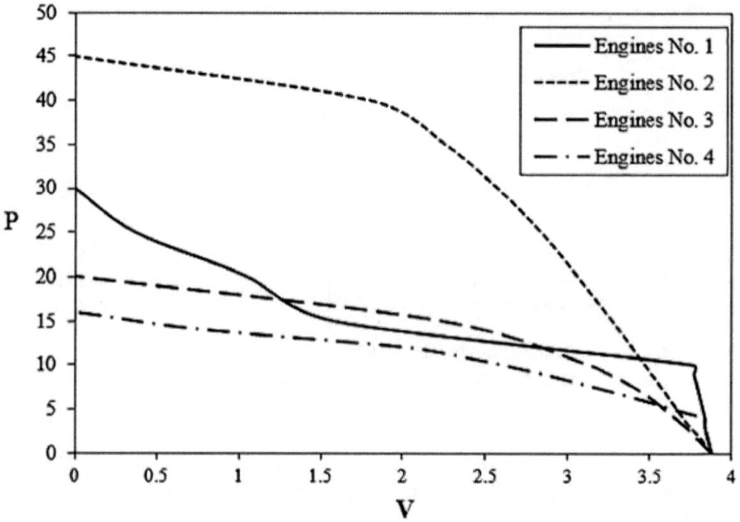

Fig. 6 The stability boundary for the fifth situation ($X_{e1} = 0.2, X_{e2} = 0.4, X_{e3} = 0.6, X_{e4} = 0.8, Y_{e(1,2,3,4)} = Z_{e(1,2,3,4)} = 0$)

4 Conclusion

In present work, in order to assess the effects of distributed electric propulsion on aeroelastic behavior of a box wing configuration, the governing equations are derived by Hamilton's variational principle. Four electric motors are considered on each front and rear wings. Unsteady model based on Wagner function is utilized to apply the aerodynamic forces and moments. The effects of the electric motors working arrangement on the box wing flutter stability boundary are evaluated for five situations. One of the advantages of distributed electric propulsion is that the aircraft pilot can shut down a number of engines in some flight conditions, such as cruise mode, and save battery power. The numerical results show that the electric motors working arrangement has valuable influence on the stability boundary of the box wing.

References

Amoozegar MR, Irani S (2012) Bending-torsional flutter of a composite rectangular cantilever wing subjected to engine thrust. J Adv Mater Res 463–464:1568–1572

Amoozgar MR, Friswell MI, Fazelzadeh SA, Haddad Khodaparast H, Mazidi A, Cooper JE (2021) Aeroelastic stability analysis of electric aircraft wings with distributed electric propulsors. J MDPI Aerosp 8(4):100

Bamberger J (2021) Wing flutter suppression via distributed electric propulsion, honors baccalaureate of science. Oregon State University, Corvallis

Fazelzadeh SA, Mazidi A (2010) Flutter of a swept aircraft wing with a powered engine. J Aerosp Eng 23(4):243–250

Fazelzadeh SA, Mazidi A, Kalantari H (2009) Bending-torsional flutter of wings with an attached mass subjected to a follower force. J Sound Vib 323:148–162

Feldt WT, Herrmann G (1974) Bending-torsional flutter of a cantilevered wing containing a tip mass and subjected to a transverse follower force. J Franklin Inst 297(6):467–478

Frediani A, Cipolla V, Rizzo E (2012) The PrandtlPlane configuration: overview on possible applications to civil aviation. In: Buttazzo G, Frediani A (eds) Variational analysis and aerospace engineering, mathematical challenges for aerospace design. Springer, Berlin

Ghasemikaram AH, Mazidi A, Fazelzadeh SA, Scholz D (2022) Flutter analysis of a 3-d box-wing aircraft configuration. Int J Struct Stab Dyn 22(2):1–24. https://doi.org/10.1142/S021945542250016X

Hodges DH, Patil MJ, Seungmook C (2002) Effect of thrust on bending-torsion flutter of wings. J Aircr 39(2):371–376

Mazidi A, Fazelzadeh SA (2013) Aeroelastic modelling and flutter prediction of swept wings carrying twin powered engines. J Aerosp 26:586–593

Memmolo V, Marano AD, Maio L, Nicolosi F, Marulo F (2022) Aeroelastic assessment of distributed electric propulsion wings. In: IOP conference series: material science and engineering, 11th EASN, Vancouver

Schiktanz D (2011) Conceptual design of a medium range box wing aircraft, M.Sc. Thesis, Automotive and Aeronautical Engineering, Hamburg University of Applied Sciences, Hamburg

Xu J, Ma X (2014) Effects of parameters on flutter of a wing with an external store. J Adv Mater Res 853:453–459

Zafari E, Mazidi A, Jalili MM (2018) Analytical nonlinear flutter and sensitivity analysis of aircraft wings subjected to a transverse follower force. Proc Inst Mech Eng: Part G J Aerosp Eng 233(4): 1–13

Zuo QH, Schreyer HL (1996) Flutter and divergence instability of non-conservative beams and plates. J Solids Struct 33(9):1355–1367

Force Attenuation Properties of Multilayer Polyurethane and 3D Fabric Composites

Mohammad Rauf Sheikhi and Selim Gürgen

Nomenclature

PU	Polyurethane
UAM	Urban Air Mobility
VTOL	Vertical Take-off and Landing
WKSF	Warp-knitted spacer fabrics

1 Introduction

Both the automotive and aviation industries have created design tools to improve occupant and internal part safety. Maximum reaction forces are important for assessing structural protective behavior, and these forces should be kept at lower levels to reduce damage at the back face. While providing adequate total energy-absorption capacity during the large deformation process, the peak force of the energy-absorbing structures/materials under impact must be kept below the threshold that would cause damage or injury; the reaction force should remain constant or nearly constant to avoid an excessively high rate of retardation (Lu and Yu 2003). Polyurethane is a form of polymer composed of organic units joined by urethane bonds. Unlike other conventional polymers like polyethylene and polystyrene, polyurethane can be manufactured from a wide range of basic materials.

M. R. Sheikhi (✉)
Faculty of Aeronautics and Astronautics, Eskişehir Technical University, Eskişehir, Turkey
e-mail: mohammadraufsheikhi@ogr.eskisehir.edu.tr

S. Gürgen
Department of Aeronautical Engineering, Eskişehir Osmangazi University, Eskişehir, Turkey
e-mail: sgurgen@ogu.edu.tr

T. H. Karakoc et al. (eds.), *Emerging Trends in Electric Aviation*, Sustainable Aviation, https://doi.org/10.1007/978-3-031-37299-5_6

Polyurethanes with a wide range of chemical structures can be employed in a variety of applications because of this chemical diversity (Allen et al. 2015). Because of their unique structure and excellent performance, three-dimensional spacer fabrics have a wide range of applications in aerospace, military, construction, shipbuilding, transportation, medicine, and other fields (Kyosev 2019). They can be made in a variety of ways, including knitting, weaving, braiding, and non-woven. WKSFs have been identified as one of the most promising materials due to their low cost, high production efficiency, and versatility. The deformation capacity of spacer fabrics is strongly related to the deformation of the spacer yarn used in the fabric. As the number of spacer yarn threads in the structure increases, so does the impact strength of spacer textiles. The spacer yarn connectors are thick and long, and the outer layer is finished (Kyosev 2019). All these properties make spacer textiles suitable for use as a protective material in impact-resistant applications (Palani Rajan et al. 2016). Due to their cellular-based bonded structures, both WKSF and PU materials have excellent anti-shock and compressibility properties (Lu et al. 2013; Sheikhi and Gürgen 2022). As a result, many researchers have devoted their time to studying the properties and influencing factors of WKSFs using multi-angle analysis. Many scholars conduct extensive research on the compressive behavior and mechanism of WKSFs under static loading (Liu and Hu 2016). WKSF and PU materials are easy to manufacture, install, and maintain and cost-effective. After impact force, a slight deformation occurs in their structure which is an important factor for aerospace and automotive structures. As a result, all protective structures must operate within these economic and usability parameters. This is especially true for energy-absorbing devices, which are typically one-shot items, meaning they are discarded and replaced after becoming deformed.

2 Experimental Details

PU layers were provided by ESPOL Sünger Inc., while WKSF layers were provided by Ames Europe Inc. in this study. By stacking four layers of 5 mm thick PU and WKSF laminates, a total thickness of 20 mm was achieved in the final design. Multilayers were initially offered in two designs: full WKSF and full PU. In the third design, two layers of WKSF were placed between the PU layers, and finally, the thickness and weight of all three designs were the same. In a drop tower system, the composites were subjected to low-velocity impact testing. Before the impacts, a hemispherical impactor was filled with 15 J. In the testing, the drop height was 0.2 meters. A load cell was located under the specimens to trace the back face force curves. Figure 1 shows the impact test system and multilayer composite designs of PU and WKSF.

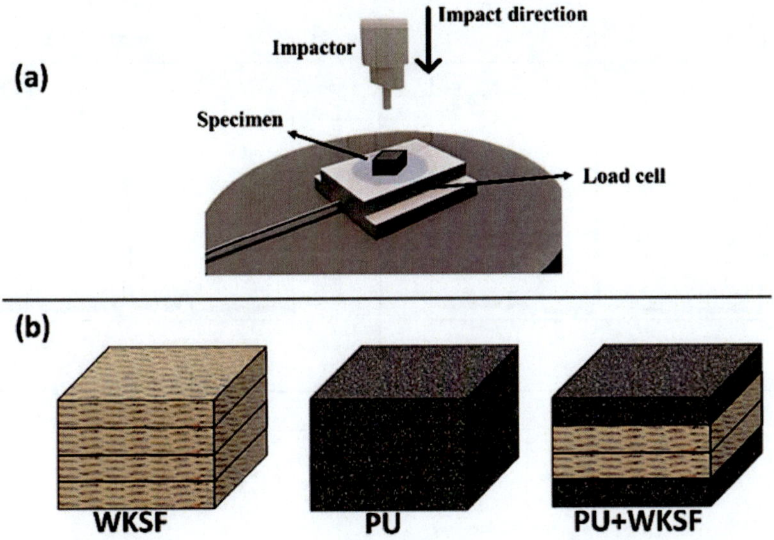

Fig. 1 (**a**) Impact test system (**b**) multilayer composite designs for impact tests

3 Results and Discussion

Figure 2 depicts the impact reaction force (force attenuation value) graphs at the composites' rear face. The curves demonstrate that the combination of WKSF and PU has a lower response force. The rear face force for this target configuration is 286 N, while the complete WKSF sample is 344 N and the whole PU is 443 N. Based on these findings, it is clear that the combination of WKSF and PU is more effective than other samples in that it absorbs the majority of the impact energy and reflects the remainder to the back face. Due to the three-dimensional structure of the WKSF layers, when these layers are placed between the PU layers, they direct the impact force in the lateral directions. On the other hand, placing WKSF layers on top of the sample cannot be effective and due to the excellent compressibility of these materials between the layers can be more effective.

4 Conclusions

In this study, we investigated the impact reaction force attenuation properties of multilayer composites consisting of WKSF and PU. We measured the factor of reaction force after impact on the samples by placing a load cell under the prepared samples and testing it with a low-velocity impact system. Although the full-WKSF sample has less reaction force than the full-PU sample, when the WKSF layers are placed between the PU layers, the reaction force shows a sharp decrease and is 35.4% better than the full-PU sample, and compared to the WKSF sample, it shows

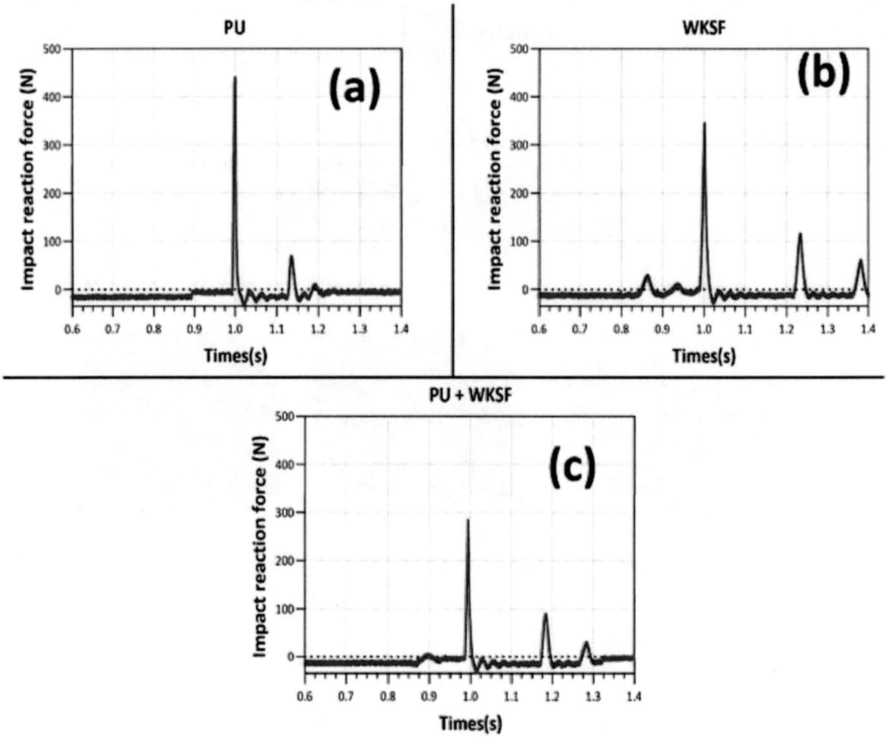

Fig. 2 Reaction force curves (force attenuation values) for (**a**) PU, (**b**) WKSF, and (**c**) PU + WKSF

17% less reaction force. WKSF and PU can be used as core layers in the structures of VTOL or the automotive industry where the reaction force factor can cause irreparable damage to the crew or their internal components.

References

Allen T, Shepherd J, Hewage T, Senior T, Foster L, Alderson A (2015) Low-kinetic energy impact response of auxetic and conventional open-cell polyurethane foams. Phys Status Solidi B 252: 1631–1639

Kyosev Y (2019) Warp knitted fabrics construction. CRC Press

Liu Y, Hu H (2016) Compressive mechanics of warp-knitted spacer fabrics. Part I: a constitutive model. Text Res J 86:3–12

Lu G, Yu T (2003) Energy absorption of structures and materials. Elsevier

Lu Z, Jing X, Sun B, Gu B (2013) Compressive behaviors of warp-knitted spacer fabrics impregnated with shear thickening fluid. Compos Sci Technol 88:184–189

Palani Rajan T, Ramakrishnan G, Kandhavadivu P (2016) Permeability and impact properties of warp-knitted spacer fabrics for protective application. J Textile Inst 107:1079–1088

Sheikhi MR, Gürgen S (2022) Anti-impact design of multi-layer composites enhanced by shear thickening fluid. Compos Struct 279:114797

Transport Operators Total Load Comparison by Analytical Hierarchy Process (AHP)

Omar Alharasees and Utku Kale

Nomenclature

AHP	Analytical hierarchy process
CR	Consistency ratio
MCDM	Multi-criteria decision-making
PCM	Pairwise comparison matrix

1 Introduction

The sharp increase in articles on sustainable mobility across all modes of transportation shows that there is a robust discussion going on in this sector, particularly over the role of the human aspect in highly automated systems. However, there is still a dearth of empirical research that analyzes the duty of operators and the development of their responsibilities and abilities in the dynamic transport environment which are crucial for reaching the aims of sustainable transportation.

The human factor is considered the most critical element affecting transport safety. Identification and classification of operators' behavior are crucial for minimizing the rate of severe accidents and injuries in the transportation network all modes considered. It is necessary to develop the environment and transportation networks for future operators to account for numerous psychological aspects, human factors, and operator loads. Evaluation of the performance of the vehicle operators, including situation awareness, decision-making, and operator load, is necessary as

O. Alharasees (✉) · U. Kale
Department of Aeronautics and Naval Architecture, Budapest University of Technology and Economics, Budapest, Hungary
e-mail: oalharasees@edu.bme.hu; kale.utku@kjk.bme.hu

T. H. Karakoc et al. (eds.), *Emerging Trends in Electric Aviation*, Sustainable Aviation, https://doi.org/10.1007/978-3-031-37299-5_7

the transportation system evolved to be more complicated. For a clear sense of automation and load balancing in complicated dynamic systems, vehicle operators (drivers and pilots) require reliable models and concepts in advanced transport systems. Performance evaluation for vehicle operators is more difficult in this highly complex and dynamic environment. However, accidents continue to happen despite all the automated improvements in transportation technology.

The future operator environment (drivers and pilots) in highly automated systems needs to be redesigned by considering various psychological parameters, human factors, and operator loads (Kale et al. 2017; Jankovics and Kale 2019). Therefore, this study focuses on evaluating operators' total loads using the analytical hierarchy process.

A reliable tactic for handling complex decisions is the Analytic Hierarchy Process (AHP), a well-recognized "Multicriteria Decision-Making (MCDM)" tool for various optimization ranking methods (Saaty 2008). Using this approach, decision-makers will establish priorities and select the best approach (Saaty 2002). The AHP combines the results of pairwise comparisons to break down complex judgments into their subjective and objective aspects. A helpful method for evaluating the consistency of the decision maker's judgments is also included in the AHP, which helps to remove decision-making bias.

The AHP is a theory of comparative measurement on ultimate scales of both direct and indirect criteria based on the perception of a familiar involved participant as well as actual data and relevant information. The main goal of the AHP's mathematics is to determine how to quantify entities by assessing and weighting the crucial elements. The AHP is used to make multi-objective, multi-criteria, and multi-party selections and decisions, particularly in the engineering sector since it needs to build a hierarchal model based on the current components of the investigated purpose (Nakagawa and Sekitani 2004). Rather than allocating a score based on a person's subjective assessment, the assessments often made in qualitative terms are presented numerically to achieve compromises among the different intangible aims and criteria (Saaty 2008). Finally, incorporating repeated and extensive experiences would flow into a system of priorities in order to deal with the challenges.

AHP was used in the transportation literature in a large number of prior research. In terms of environmentally sustainable factors, Yedla and Shrestha (2003) studied and chose the best alternative in the Delhi transportation system (Yedla and Shrestha 2003). Bruno et al. evaluated planes to preserve scheduled selections, revealing that cabin luggage compartment space is the best element (Bruno et al. 2015). Policy and reliability were found by Chao and Kao to be essential criteria for service quality (Chao and Kao 2015). Rezaei et al. (2014) measured and identified the supplier in the airline retail business; the outcome of this research indicated that economic stability is a considerable standard in supplier selection (Rezaei et al. 2014). Chen (2014) utilized AHP Technique to evaluate the impact of weighting the technical aspects in aviation safety (Chen et al. 2014).

Some researches focused on the drivers and pilots in the transport systems, like investigating the vehicle drivers behavior criteria for evolution of sustainable traffic safety (Farooq et al. 2019; Moslem et al. 2020). Cen, Jiayin proposed a system architecture that aims to combine vehicle driver background data and driving data, in order to analyze and evaluate driving behaviors (Cen et al. 2017). Oktal H and Onrat A. utilized AHP for indicating the important aspects in airline pilots' candidates' selection (Oktal and Onrat 2020). Havle and Kılıç (Havle and Kılıç 2019) found and assessed the considerations that affect navigation errors in the North Atlantic Region by combining a Fuzzy Analytic Hierarchy Process (FAHP) into the Human Factors Analysis and Classifying System structure (Havle and Kılıç 2019). Kilic and Ucler (Kilic and Ucler 2019) employed AHP Method to assess stress factors among student pilots (Kilic and Ucler 2019).

The research aims to evaluate the elements that influence and shape the total loads of transport operators (drivers and pilots) and compare the difference between the two systems critical elements. The current study examines the preferences of the four operator categories (less skilled pilots, skilled pilots, less skilled vehicle drivers, and skilled vehicle drivers) based on the primary criteria. In order to create a general hierarchical model, the Analytic Hierarchy Process (AHP) is employed in this research. These decision-making models are primarily built on three layers in order to develop evaluator preference loads for (i) the assessment procedure, (ii) preventing complication, and (iii) lacking information from other AHP functions. In this study, the Saaty Scale was utilized for scoring to depict lost data utilizing matrices that could be computed using a particular technique.

2 Method

Decision alternatives or sub-criteria must be picked or selected according to their attributes in order to use the MCDM approach. A predetermined, constrained number of decision alternatives is given in MCDM situations. In the MCDM process, there are three steps: sorting, ranking, and scoring.

The Analytic Hierarchy Process (AHP), a famous Multi-Criteria Decision-Making (MCDM) tool, is the principal study method used to examine the key and fundamental aspects of pilots and vehicle drivers.

The current authors created a two-level hierarchy model containing the five main categories of transport operators loads: (i) workload is the work performed by vehicle operator in a given time, it depends on human factors, skills, knowledge, practice, etc.; (ii) task load is the degree of difficulty and challenge when carrying out a task, which alters by the degree of difficulty, traffic demand, traffic regulations, etc.; (iii) information load is the amount of information and data received from the highly automated systems, which is affected by the level of technology and other aspects; (iv) communication load is the level of awareness and understanding between operators and is highly altered by cultural norms, social relations, etc.; and (v) mental load is the physical and psycho-physiological situation of the operators during operation and highly depends on the level of stress, performance action, etc.

The hierarchical model for the total loads of operators with the components of each level is shown in Fig. 1.

Because the AHP employs the pairwise comparison matrices (PCM) properties, the choice of decision-makers between predetermined pairs of options demonstrates the significance and priority of a certain element over another based on a scale (see Table 1). The matrix of pairwise comparisons (see eq. 1) A = [aij] represents the strength of the decision-maker's preference between individual pairs of alternatives (Ai versus Aj, for all i, j = 1, 2,. . ., n). The pairwise comparison matrix can be given as follows (Eq. 1):

$$A = \left[a_{ij}\right] = \begin{bmatrix} 1 & a_{12} & . & . & a_{1j} & . & . & a_{1n} \\ \frac{1}{a_{12}} & 1 & . & . & a_{2j} & . & . & a_{2n} \\ . & & . & & & & & . \\ . & & & . & & & & . \\ \frac{1}{a_{1j}} & \frac{1}{a_{2j}} & . & . & a_{ij} & . & . & a_{in} \\ . & . & & & & & . & . \\ . & . & & & & & . & . \\ \frac{1}{a_{1n}} & \frac{1}{a_{2n}} & . & . & \frac{1}{a_{in}} & . & . & 1 \end{bmatrix} \quad (1)$$

The geometric mean of each group was calculated in the pairwise comparison matrices to assess and rank the impact of each model element at the same level. Given that the majority of experience matrices are inconsistent, the matrix consistency ratio (CR) should be smaller than 0.1.

3 Questionnaire

In this research, an online AHP-based survey was designed and performed for vehicle operators (drivers and pilots) a total of 40 participant in four groups of operators, focusing on the operators' major characteristics from various perspectives. The purpose of the questionnaire is to quantify the most important issues as seen through the eyes of the operators, based on their experience and knowledge.

The questionnaire was created based on vehicle operators (drivers and pilots) in this research. There were 40 participants (6 females, M = 29.4, SD = 7.3). The participants were arranged into four groups: (i) less-skilled pilots (N = 10, 1 female, M = 21.7, SD = 1.8), (ii) skilled pilots (N = 10, 2 females, M = 35.4, SD =2.7), (iii) less skilled vehicle drivers (N = 10, 3 Females, M = 25,0.6 SD = 4), and (iv) skilled vehicle drivers (N = 10, M = 34.8, SD = 7.1).

Fig. 1 General hierarchical model of the operators' total load

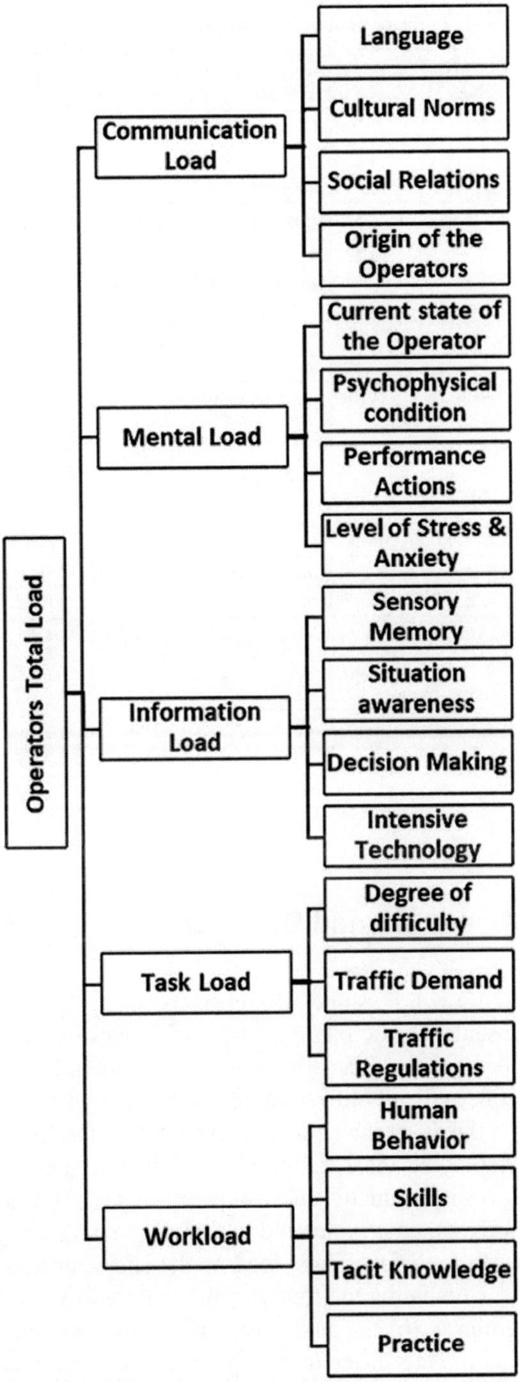

Table 1 Saaty fundamental scale

Numerical values	Verbal scale	Explanation
1	Equal importance of both elements	Two elements contribute equally
3	Moderate importance of one element over another	Experience and judgment favor one element over another
5	Strong importance of one element over another	An element is strongly favored
7	Very strong importance of one element over another	An element is very strongly dominant
9	Extreme importance of one element over another	An element is favored by at least an order of magnitude
2, 4, 6, 8	Intermediate values	Used to compromise between two judgments

Table 2 Less skilled pilots PCM for the first level

Less skilled pilots						
Operators total loads	Communication load	Mental load	Information load	Task load	Workload	Weights
Communication load	1	4.45	4.51	2.89	1.14	39.35%
Mental load	0.22	1	2.62	2.04	0.48	15.48%
Information load	0.22	0.38	1	3.16	0.43	11.71%
Task load	0.35	0.49	0.32	1	0.49	8.38%
Workload	0.88	2.09	2.32	2.05	1	25.08%
CR = 0.096	Sum=					100%

4 Results and Discussions

There will be some differences between the groups' overviews after analyzing and visualizing the participants' preferences in the model; AHP method will highlight the critical characteristics based on pairwise comparisons. The responses have been collected and utilized by using the geometric mean.

Based on the collected responses of the four groups of vehicle operators and by employing the AHP process, evaluating, and weighing the characteristics in each level individually, the following tables (Tables 2, 3, 4, 5) show the aspects (the weights, final score, and consistency ratio) which have been computed for the first level in the operators' load model characteristics from each group:

Combining the four groups' preferences would show the variations between the groups, which could raise due to the experience level and the type of the job. Comparing different groups of participants would make it easier to evaluate and weigh various individual aspects of transport operators' total loads from other overviews.

Table 3 Skilled pilots PCM for the first level

Skilled pilots						
Operators Total loads	Communication load	Mental load	Information load	Task load	Workload	Weights
Communication load	1	3.98	3.60	4.13	0.82	36.33%
Mental load	0.25	1	0.95	1.60	0.35	10.91%
Information load	0.28	1.05	1	3.73	0.45	14.95%
Task load	0.24	0.63	0.27	1	0.39	7.46%
Workload	1.22	2.84	2.23	2.58	1	30.34%
CR = 0.047	Sum=					100%

Table 4 Less skilled vehicle drivers PCM for the first level

Less skilled vehicle drivers						
Operators total loads	Communication load	Mental load	Information load	Task load	Workload	Weights
Communication load	1	1.90	0.37	1.34	0.21	12.82%
Mental load	0.53	1	1.03	1.85	0.35	13.93%
Information load	2.69	0.97	1	2.53	0.29	19.74%
Task load	0.75	0.54	0.40	1	0.51	10.59%
Workload	4.84	2.87	3.41	1.95	1	42.92%
CR = 0.095	Sum=					100%

Table 5 Skilled vehicle drivers PCM for the first level

Skilled vehicle drivers						
Operators total loads	Communication load	Mental load	Information load	Task load	Workload	Weights
Communication load	1	1.30	0.28	0.87	0.29	11.30%
Mental load	0.77	1	1.08	1.57	0.36	15.67%
Information load	3.60	0.92	1	2.44	0.44	23.77%
Task load	1.15	0.64	0.41	1	0.67	13.11%
Workload	3.42	2.80	2.29	1.50	1	36.15%
CR = 0.0737	Sum=					100%

The survey highlighted a huge discrepancy between the pilot's groups and the vehicle driver's groups particularly between the communication loads and the workloads which are the critical factors in the model in the first hierarchy level, as illustrated by Fig. 2.

A noticeable fluctuation of the opinions is clear between the groups in the third and fourth critical factors (information and mental loads) within the first level of the model.

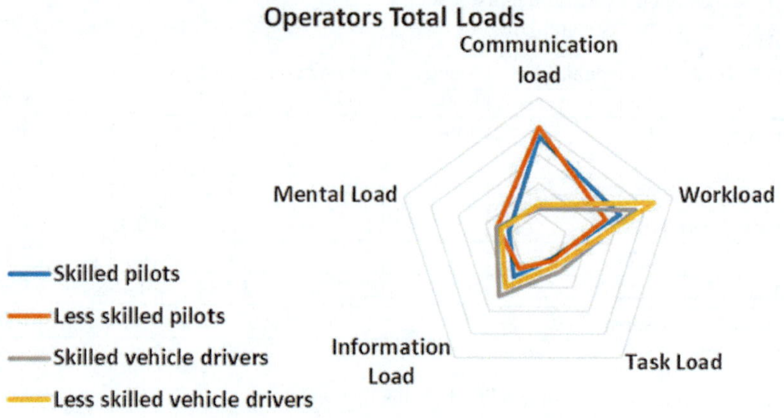

Fig. 2 The total load of the transport operators

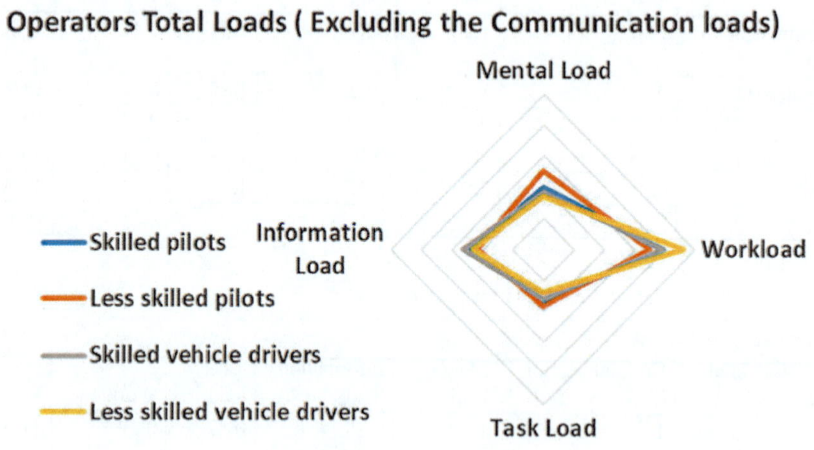

Fig. 3 The total load of the transport operators (excluding the communication load)

Communication load is a significant element in the aviation system since the operators must communicate accurately and successfully to assure the safety of the flight; however, the situation is different in land transport since the communication load is not critical in the current systems at the same level as in the air transport that is why excluding the communication load would give a more clear and precise comparison between the groups as shown in Fig. 3.

Based on the participant's opinions, the information load is the most critical issue for the futuristics transport operator's environment after the workload, especially with introducing the autonomous systems in the transport systems.

5 Conclusion

The findings demonstrated a priority ranking and scaling of the operators' total loads within each level, which is a great indicator of the significant elements. To better understand the futuristic operators' environment and manage critical scenarios, employing multi-criteria procedures, specifically AHP, illustrated a critical role. The inconsistencies between the perspectives are shown using quantitative and qualitative criteria using the traditional, classic, and simplified analytical hierarchical process (AHP) technique for decision-making.

The results of this survey were based on a total of 40 participants from four groups of transport operators (less skilled pilots, skilled pilots, less skilled vehicle drivers, and skilled vehicle drivers); it should be mentioned that the results could change if more participants and more groups are included.

The results show that the workload load and information load play a dominant role in the operator total loads model from all participants, followed by the mental load of operators.

References

Bruno G, Esposito E, Genovese A (2015) A model for aircraft evaluation to support strategic decisions. Expert Syst Appl 42(13):5580–5590. https://doi.org/10.1016/J.ESWA.2015.02.054

Cen J et al. (2017) A system design for driving behavior analysis and assessment. In: Proceedings – 2016 ieee international conference on internet of things; IEEE Green computing and communications; IEEE cyber, physical, and social computing; IEEE smart data, ithings-greencom-cpscom-smart data 2016. Institute of Electrical and Electronics Engineers Inc., pp 882–887. doi: https://doi.org/10.1109/ITHINGS-GREENCOM-CPSCOM-SMARTDATA.2016.182

Chao CC, Kao KT (2015) Selection of strategic cargo alliance by airlines. J Air Transp Manage 43: 29–36. https://doi.org/10.1016/J.JAIRTRAMAN.2015.01.004

Chen CJ, Yang SM, Chang SC (2014) A model integrating fuzzy AHP with QFD for assessing technical factors in aviation safety. Int J Mach Learn Cybern 5(5):761–774. https://doi.org/10.1007/S13042-013-0169-1/TABLES/8

Farooq D, Moslem S, Duleba S (2019) Evaluation of driver behavior criteria for evolution of sustainable traffic safety. Sustainability 11(11):3142. https://doi.org/10.3390/SU11113142

Havle CA, Kılıç B (2019) A hybrid approach based on the fuzzy AHP and HFACS framework for identifying and analyzing gross navigation errors during transatlantic flights. J Air Transp Manag 76:21–30. https://doi.org/10.1016/J.JAIRTRAMAN.2019.02.005

Jankovics I, Kale U (2019) Developing the pilots' load measuring system. Aircr Eng Aerosp Technol 91(2). https://doi.org/10.1108/AEAT-01-2018-0080

Kale U, Tekbas, MB, Rohacs J, Rohacs D (2017) System supporting the operators supervising with vehicle and transport control. In Peter DT (ed) IFFK 2017. Budapest, pp 101–108

Kilic B, Ucler C (2019) Stress among ab-initio pilots: a model of contributing factors by AHP. J Air Transp Manage 80:101706. https://doi.org/10.1016/J.JAIRTRAMAN.2019.101706

Moslem S et al (2020) Application of the AHP-BWM model for evaluating driver behavior factors related to road safety: a case study for Budapest. Symmetry 12(2):243. https://doi.org/10.3390/SYM12020243

Nakagawa T, Sekitani K (2004) A use of analytic network process for supply chain management. Asia Pac Manag Rev 9:783–800

Oktal H, Onrat A (2020) Analytic hierarchy process–based selection method for airline pilot candidates. Int J Aerosp Psychol 30(3–4):268–281. https://doi.org/10.1080/24721840.2020.1816469

Rezaei J, Fahim PBM, Tavasszy L (2014) Supplier selection in the airline retail industry using a funnel methodology: conjunctive screening method and fuzzy AHP. Expert Syst Appl 41(18):8165–8179. https://doi.org/10.1016/J.ESWA.2014.07.005

Saaty TL (2002) Decision making with the analytic hierarchy process. Sci Iran 9(3). https://doi.org/10.1504/ijssci.2008.017590

Saaty TL (2008) Decision making with the analytic hierarchy process. Int J Serv Sci 1(1):83–98

Yedla S, Shrestha RM (2003) Multi-criteria approach for the selection of alternative options for environmentally sustainable transport system in Delhi. Transp Res A Policy Pract 37(8):717–729. https://doi.org/10.1016/S0965-8564(03)00027-2

Analysis of Safety Risks Related to Alternative Aviation Fuels

Martina Koščáková, Samer Al-Rabeei, Peter Korba, and Utku Kale

1 Introduction

The issues related to the impact of air transport on the environment, especially the effects of the Greenhouse Gas emissions produced, are currently the main focus of many research projects and regulations. Although it is highly important to replace conventional aviation fuels with suitable alternative fuel, it is challenging, as characteristics of new sustainable aviation fuels should be similar or even better than those of conventional fuels. The key aspects are safety and efficiency under extreme temperature changes. Moreover, it is preferable that new aviation fuels be implemented to already existing infrastructure without significant changes to the aircraft design. Therefore, development of just right alternative aviation fuel is a daunting task.

2 Alternative Aviation Fuels

Sustainable aviation fuels are defined as bio-based jet fuels designed to have lower emissions compared to the conventional ones, while at the same time preventing other negative sustainability impacts (EASA 2022). These fuels are produced from

M. Koščáková (✉) · S. Al-Rabeei · P. Korba
Faculty of Aeronautics, Technical university of Košice, Košice, Slovakia
e-mail: martina.koscakova@tuke.sk; samer.al-rabeei@tuke.sk; peter.korba@tuke.sk

U. Kale
Department of Aeronautics and Naval Architecture, Budapest University of Technology and Economics (BME), Budapest, Hungary
e-mail: kale.utku@kjk.bme.hu

T. H. Karakoc et al. (eds.), *Emerging Trends in Electric Aviation*, Sustainable Aviation, https://doi.org/10.1007/978-3-031-37299-5_8

bio-based feedstock that include oils, fats and greases, residues, and end-of life products, as well as fossil waste (RSB 2022). The main requirements for the feedstock sources are as follows (Sustainable Aviation 2020):

- Are manufactured from biomass or recycled carbon.
- Meet rigorous sustainability standards regarding land, water, and energy use.
- Avoid direct and indirect land use change impacts, such as tropical deforestation.
- Do not displace or compete with food crops.
- Produce a positive socio-economic impact.
- Display minimal impact on biodiversity and conservation values.
- Have been assessed and certified by an appropriate sustainability standard.

The rate of GHG emissions saving from alternative sources is considerable, however varies significantly depending on the raw materials utilized for its production. For the electricity-based solutions, the savings could be up to 95%, while solutions based on crops are expected to provide the reduction of 65% of GHG emissions, and biomass wastes can be found somewhere within this range (International Maritime Organization 2013).

3 Risks

To ensure the safety of the flight, all the fuels within commercial airplanes must comply criteria established by engine manufacturers and must be approved by the safety agencies. Certification of new aviation fuels is within the scope of the ASTM International Aviation Fuels Subcommittee (Bauen and Nattrass 2017).

3.1 Compatibility with Conventional Fuel

Alternative aviation fuels need to fulfil several requirements concerning their compatibility with conventional jet fuels. New sustainable fuels need to be highly resistant to number of different circumstances, must have substantial performance, and must be guaranteed as safe to use. Among the most important issues related to biofuels are thermal and storage stability, characteristics at low temperatures, combustion characteristics, and presence of micronutrients (Hari et al. 2015). One of the key issues that needs to be regarded when considering utilization of alternative aviation fuels is the fuel energy density. While bio-jet fuels have almost identical density relative to conventional fuels, hydrogen's volumetric energy density is lower (International Maritime Organization 2020).

3.2 *Combustion Properties*

It is vital to reliably know fuel's combustion characteristics over a relevant parameter range, such as temperature, pressure, composition, and fuel-air ratio in order to prevent flashback or self-ignition of the fuel in the operating conditions, which could result in severe damage to the burner and the combustion chamber. The laminar flame speed has a direct impact on the flame length, and therefore, if it significantly differs from the conventional fuel, it could result in the damage to the jet turbine. If it is too low, there is an increased risk of a blowout; too high can result in flashback in premixed systems. Another factor is the ignition delay time that indicates the stability of combustion. Too low ignition delay time may result in flashback; with too high, there is the risk that ignition or re-ignition of the fuel may not be possible (Braun-Unkhoff and Riedel 2015).

3.3 *Effects on Components*

Study was conducted at the Faculty of Aeronautics of Technical university of Košice (Čerňan et al. 2017) to determine the risks related to the use of combination of Jet A-1 aviation kerosene and FAME biofuels. From the study, it was concluded that the addition of FAME to the mixture reduces the strength of rubber seals by reducing its hardness through the impact of rapeseed oil fatty acid methyl esters. Moreover, bio-fuels are more pronounce to the microbial contamination. According to several studies, there is a higher rate of microbial induced fuel degradation and higher rate of microbial induced corrosion of the alternative fuel system components relative to fossil-based fuels (Shkilniuk et al. 2019). Another area of concern is the absence of aromatics in the alternative fuels, which causes them to have lower than minimum density, and can cause them to make elastomers shrink and therefore may cause fuel leaks (Global Aviation 2006).

4 Risk Assessment

Hazard	Consequence	Severity
High freezing point	Engine non-functioning	Red
Flashback	Damage to the burner and combustion chamber	Red
Self-ignition		Red
Laminar flame	Damage to the jet turbine	Red
Ignition-delay time	Flashback, problems with ignition and re-ignition	Red

(continued)

Hazard	Consequence	Severity
Microbial induced fuel degradation	Metal degradation	Orange
Microbial induced corrosion of fuel system components		Orange
Cause elastomers to shrink	Fuel leaks	Orange
Reduced strength of rubber seals		Orange

Through the research, we were able to identify nine of the key risks related to the use of alternative fuels in aviation. For their assessment, we have not used the conventional risk assessment matrix, but rather the slimmer version, taking into account only severity not probability of occurrence of the hazard. This method was chosen due to the fact that alternative aviation fuels are under development and many of the potential risks are expected to be eliminated before their wider implementation.

As the most severe (marked red) were identified risks related to the functioning of the engine. Those are risks that could cause direct damage to the jet engine during the flight and therefore jeopardize the safety of the crew and passengers. On the other hand, as less severe (marked orange) were identified risks related to the fuel system components, which effect is not immediate.

5 Conclusion

Aviation industry is for years focused on the development of alternative aviation fuels for multiple reasons – performance, effectivity, financial, environmental, or sustainable supply. This paper provided overview of the most important pathways for the production of alternative aviation fuels. Among the most significant risks identified are issues related to the storage, performance under extreme temperatures, and fuels´ effects on the aircraft components, which may result in the compromised safety of air travel. Moreover, it can be said that fuels´ characteristics are not the only limitation to their wider implementation. Development of alternative aviation fuels is limited by the financial constraints, as well as great competition of other means of transport, with less demanding conditions and safety requirements, and other industries. Even though new types of fuels could significantly reduce GHG emissions and thus contribute to the mitigation of climate change, it can be expected that their use will remain limited for the upcoming years.

References

Aviation S (2020) Sustainable aviation fuels road-map. https://www.sustainableaviation.co.uk/wp-content/uploads/2020/02/SustainableAviation_FuelReport_20200231.pdf

Bauen A, Nattrass L (2017) Sustainable Aviation biofuels: scenarios for deployment. Biokerosene. https://doi.org/10.1007/978-3-662-53065-8_27

Braun-Unkhoff M, Riedel U (2015) Alternative fuels in aviation. CEAS Aeronaut J 6:83–93. https://doi.org/10.1016/j.rser.2014.10.095

Čerňan J et al (2017) Safety risks of biofuel utilization in aircraft operations. Transp Res Procedia 28:141–148. https://doi.org/10.1016/j.trpro.2017.12.179

EASA (2022) Sustainable aviation fuels. https://www.easa.europa.eu/eaer/climate-change/sustainable-aviation-fuels

Global Aviation (2006) Alternative jet fuels. https://www.chevron.com/-/media/chevron/operations/documents/chevron-alternative-jet-fuels.pdf

Hari TK et al (2015) Aviation biofuel from renewable resources: routes, opportunities and challenges. Renew Sust Energ Rev 42:1234–1244. https://doi.org/10.1016/j.rser.2014.10.095

International Maritime Organization (2013) Air pollution and energy efficiency: information about the application status of Tier III compliant technologies. https://www.euromot.eu/wp-content/uploads/2017/03/IMO_MARPOL_A6_NTC_information_about_the_application_status_of_Tier_III_compliant_technologies_2013-11-01.pdf

International Maritime Organization (2020) Reducing greenhouse gas emissions from ships. http://www.imo.org/en/MediaCentre/HotTopics/Pages/Reducing-greenhouse-gas-emissions-from-ships.aspx

RSB (2022) Alternative aviation fuels, a sustainable future is taking off. https://rsb.org/wp-content/uploads/2018/09/RSB-Alternative-Aviation-Fuels-A-Sustainable-Future-is-Taking-Off.pdf

Shkilniuk I et al. (2019) Identification and assessment of biological risk of aviation fuel supply. https://doi.org/10.18372/38233

Adding Value to Aviation Through Additive Manufacturing

Volodymyr Tymofiiv, Samer Al-Rabeei, Michal Hovanec, Peter Korba, and Utku Kale

Nomenclature

AM Additive manufacturing
CAD Computer-aided design

1 Introduction

Additive manufacturing, also known as 3D printing, is a manufacturing process in which a 3D printer creates three-dimensional objects by applying the material in layers, according to a digital 3D model of the object.

3D printing is the exact opposite of traditional mechanical production and machining methods such as milling or cutting, where the appearance of the product is shaped by removing excess material (Bindiganavile Anand et al. 2021).

V. Tymofiiv (✉) · S. Al-Rabeei · M. Hovanec · P. Korba
Faculty of Aeronautics, Technical university of Košice, Košice, Slovakia
e-mail: volodymyr.tymofiiv@tuke.sk; samer.al-rabeei@tuke.sk; michal.hovanec@tuke.sk; peter.korba@tuke.sk

U. Kale
Department of Aeronautics and Naval Architecture, Budapest University of Technology and Economics (BME), Budapest, Hungary
e-mail: kale.utku@kjk.bme.hu

T. H. Karakoc et al. (eds.), *Emerging Trends in Electric Aviation*, Sustainable Aviation, https://doi.org/10.1007/978-3-031-37299-5_9

1.1 The Use of AM in the Aviation Industry

The use of AM as a unique manufacturing tool has interested manufacturers since the technology's inception. With the development of 3D technology, this technique has opened up ever greater advantages for the industry, as a result of which it has been given attention by large companies specializing in a wide variety of fields of activity. Unsurprisingly, one of the hottest topics of discussion has been 3D printing in aviation. Indeed, in the aerospace industry, 3D printing can not only simplify but also significantly improve the manufacturing process.

Thus, additive technologies are an essential choice for a competitive enterprise. Today, most companies implementing them are looking to reduce production costs and product development time without changing the supply chain and product mix (Araújo et al. 2021).

2 Application Opportunities of Uses AM in Aviation

Additive manufacturing can be used not only as a counterpart or as a substitute for conventional manufacturing. AM has several applications that are not available in conventional manufacturing. The following is a look at the main technologies available in the AM market (Singamneni et al. 2019).

2.1 Rapid Prototyping

From the beginning, 3D printing technology has been conceived as a rapid prototyping technology. Developing a new type of product is a long and time-consuming process that requires several designs and evaluation phases before mass production begins. At one time, CAD modeling technology helped to dramatically accelerate these stages. However, the problem of transferring this model into physical form remained a bottleneck of the whole process for a long time, since it required time and the whole production process had to be adjusted to a single prototype part. Hence, the technology of layer-by-layer material formation became a logical continuation of prototype production because this allowed in the shortest possible time to produce CAD models for subsequent testing (e.g., in an aerodynamic tube). After the necessary tests, changes to the model are also made in the shortest possible time and it can be printed and tested again (Vashishtha et al. 2011).

The advantages of rapid prototyping are as follows:

- Reduced development cycle
- Design improvement
- Work with more complex designs

- Reducing prototype and production costs
- Speeding up design changes (Thoppil and Subbu 2014)

Sometimes in the process of developing a new product or creating replacement parts for an old one, it is necessary to use a process called reverse engineering.

This is usually the case for older aircraft where the technical documentation has been lost or is incorrect, so it was decided that the best method would be a scan. Also, with the termination of production by the supplier of spare parts best method would be a scan too. In such cases, additive manufacturing could be a quick resolution after getting a CAD model of the desired replacement part (López and Vila 2021).

Also in the aircraft industry, RE can be used to create a replica of an airplane after it has been fully assembled. This will allow us to carry out analyses in a virtual environment and then also check the printed model, for example in a wind tunnel.

2.2 *New Possibilities for Working with Materials*

Topological Optimization

The conventional manufacturing process of parts severely limits their geometry and specifications. New geometries such as honeycomb structures, lattices, and internal cavities can make the manufacturing process very expensive or even unrealistic for conventional manufacturing. But the development of AM allowed us to manufacture a new type of part with a completely different geometry. At this point, topological optimization technology began to develop. It is based on finite element analysis, and areas of material subjected to minor loads are removed to obtain final topologically optimized part shapes, that is, load analysis is performed on each part of the part in the CAD environment, and, if the part remains as safe as before, the parts that do not take more load are removed. Subsequently, a design is obtained in which the maximum amount of its area is involved when loads are applied (Zhu et al. 2021).

This method is primarily effective in that it greatly reduces the weight of the part being produced. This is particularly important in the aircraft industry, since weight reduction is a key factor in increasing productivity and material efficiency. Reducing weight in operation leads to fuel savings for the aircraft, which in turn also reduces CO2 emissions.

As an example, as part of Airbus' research activities, the EADS Innovation Works research group developed an optimized nacelle hinge bracket for the Airbus A320 aircraft. Originally the component was made of steel, but the research team decided to use Ti6Al4V material, and the manufacturing method chosen was additive layer manufacturing (ALM). The weight of their optimized design was reduced by 64%, saving a total of about 10 kg of weight per aircraft (Altair Engineering Inc. 2020).

Part Consolidation

In the process of building complex aerospace systems, many parts are used. This has its advantages because it is very easy to replace one part in case of damage or an upgrade. However, such systems require different types of connections or fasteners such as welding, soldering, bolts, and nuts. It also creates vulnerabilities at the connection point, complicates the assembly process, and lengthens the delivery chain (Debnath et al. 2022)

Additive manufacturing can solve this problem by making a complex part that combines all the parts we need. In this manner, we get the features and increased performance we need while getting rid of the dead spots that are only used for joining. As a result, the weight of the product itself is also reduced. In addition, it will reduce inventory, lead time, assembly line area, and supply chain pressure by increasing component productivity.

With the consolidation of materials, it was possible to combine nine individual parts of the flame tube design into only one part. This shortened the production cycle and assembly of Farsoon FS271M by 50%, as well as increased structural integrity and overall performance of the part by eliminating manual welding of joints. The post-processed flame tube was functionally tested under high-temperature reliability conditions, and its weight was also reduced by 20% (Farsoon Technologies 2022).

Composite Engineering

Composite materials, due to their mechanical, chemical, and thermal properties, have become indispensable sources of aircraft structural parts.

A perpetual pursuit among aircraft designers involves the continuous endeavor to decrease the weight of aircraft and enhance their structural integrity, all while incorporating intricate designs into the fuselage and other components. This has resulted in a demand for materials that are both lighter and stronger. To fulfill this requirement, extensive research has been conducted, leading to the development of diverse types of composite materials. These materials frequently exhibit favorable characteristics such as excellent weight-bearing capacity, heightened strength, resistance to corrosion, and superior durability.

The advantages of composites are often lost at the joint because additional paving and reinforcement are required. AM allows the production of geometrically complex three-dimensional objects and offers new possibilities for processing composite materials. It makes it easy to add or remove material and thus adapt the mechanical characteristics to local loads without additional weight.

In addition, the use of AM allows us to selectively change the type of material in the part area. Although the full potential of this aspect has not yet been realized, it is expected to be used extensively to embed sensors/chips/wires directly into the printed part (Türk et al. 2016).

2.3 AM in Aviation Supply Chains

The supply chain rationale for using additive manufacturing in the aviation industry is quite compelling because it solves some key problems: the complexity of the supplier base, rigidity of the supply chain, long lead times, losses, and others.

Standard chains are extremely long and production plants can be located in another country and on another continent. In addition, the standard process of manufacturing parts is complex and often requires a complete re-equipment of industrial lines. In the case of creating a large number of spare parts with an eye to the future, there are additional costs for the maintenance of these parts, and it is unknown when they will be useful and whether they will be useful at all. Therefore, in this case, technological obsolescence of those parts that are idle in warehouses is possible. There is a large amount of waste and extra costs. All these features reduce the quality of the delivery chain, while airlines need the necessary spare parts with the highest speed of response and order fulfillment at the lowest price.

In the case of AM, one manufacturer supplier can supply a much wider range of parts printed using different technologies and materials. One AM supplier can supply what is normally produced on a large number of injection molding lines distributed among many suppliers. In addition, spare parts production areas can be located right next to service areas or right on their territory (Materialize 2020).

In industries where very complex parts are produced in small series, 3D printing becomes an ideal solution. Using additive technology, parts with complex configurations can be produced without the need for expensive conventional equipment. This solution allows aerospace component manufacturers to make low-volume production cost-effective.

For the same reason, there is a good opportunity to create more customer-oriented production and customization. Since the production cost is no longer so strongly dependent on quantity, it is possible to easily modify the design to meet the requirements of individual customers. This can be used, for example, to customize aircraft cabin designs (Zijm et al. 2019) (Fig. 1).

3 Analysis of the Interaction Between AM and the Aircraft Manufacturing Sector

The capabilities of additive manufacturing in the aircraft industry mainly allow for advances and new solutions in three areas:

1. Optimization of product design and development
2. Flexibility in the supply chain
3. Improved performance of parts

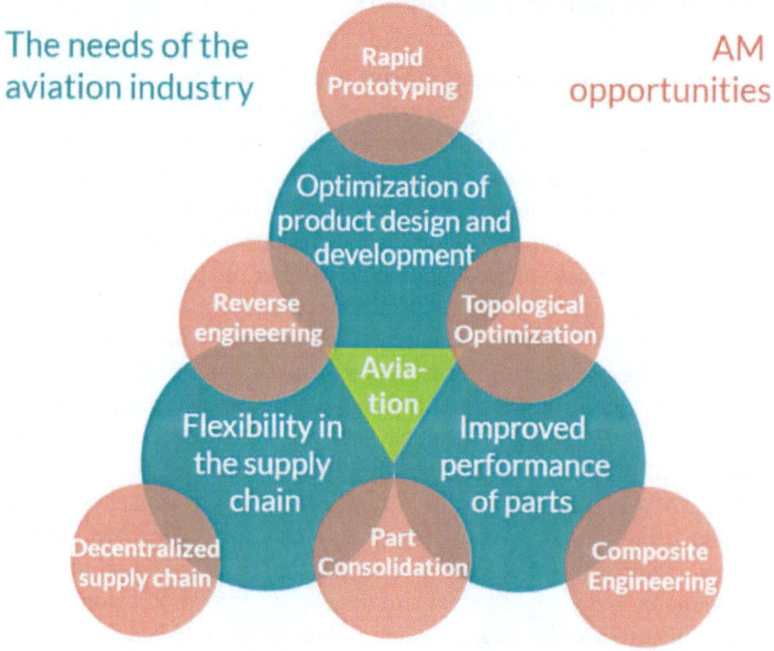

Fig. 1 Matching requirements aviation industry and capabilities AM. (Source: own elaboration)

In the planning and design phase of production, AM mainly benefits from the use of rapid prototyping, which saves a lot of time in product development, analysis, and testing. At the same time, the use of scanners and the reverse engineering process allows for greater product development capabilities, faster production of spare parts, and a new space for product analysis.

Just as importantly, using AM gives designers more freedom to design and customize parts. Technologies such as topological optimization and material consolidation allow us to model parts with geometries that cannot be produced by conventional material handling methods, but by using AM we can produce them.

As a consequence, we get parts with new geometry. Often it is possible to save a lot of weight on the part and not lose the previous characteristics or even improve them. In addition, the use of composite materials in conjunction with AM technology can improve part performance.

In addition to new design and part creation capabilities, AM also enables the optimization of the part supply chain. With a CAD model of the desired part, it can be produced on any 3D printer available for its production, without having to customize complex production for a single part. This would make it possible to dispense with fewer part companies, reduce the logistics chain, and make the whole supply chain more flexible. In addition, part consolidation technology could allow for a reduction in the number of parts needed, which could also simplify the delivery chain.

4 Conclusion

From its first use in the aviation industry, additive technologies have immediately struck people's minds with the possibilities for their further development. It became abundantly clear that the main advantages of additive manufacturing are best matched by the aviation and space technology sector.

The huge number of details in the structure of the main components of the aircraft, the requirements for the lowest possible weight and the highest possible strength of the materials, the unusual design requirements, the small number of manufactured parts (compared to other sectors), and the need for their rapid improvement have acted as a catalyst for the development of additive manufacturing.

Today, additive manufacturing also has many limitations, but already many aerospace companies use the technology, and they are not afraid to use it for the most critical parts of the aircraft, such as the engine, combustion chamber, and other parts. It is worth saying that the space for the development of this technology remains enormous and will increase with the measure of the increase in the level of technological progress.

References

Altair Engineering, Inc. (2020) Topology optimisation of an aerospace part to be produced by Additive Layer Manufacturing (ALM). https://www.messe.de/apollo/hannover_messe_2020/obs/Binary/A1003575/1003575_02137596.pdf

Araújo N, Pacheco V, Costa L (2021) Smart additive manufacturing: the path to the digital value chain. Technologies 9:88. https://doi.org/10.3390/technologies9040088

Bindiganavile Anand P, Lokesh N, Buradi A, Santhosh N (2021) A comprehensive review of emerging additive manufacturing (3D printing technology): methods, materials, applications, challenges, trends, and future potential. Mater Today Proc 52. https://doi.org/10.1016/j.matpr.2021.11.059

Debnath B, Shakur MS, Tanjum F, Rahman MA, Adnan ZH (2022) Impact of additive manufacturing on the supply chain of aerospace spare parts industry—a review. https://doi.org/10.3390/logistics6020028

Farsoon Technologies (2022) Additive manufacturing of combustion chamber flame tubes with Farsoon metal solutions. https://www.farsoon-gl.com/alcggs/additive-manufacturing-of-combustion-chamber-flame-tubes-with-farsoon-metal-solutions/

López J, Vila C (2021). An approach to reverse engineering methodology for part reconstruction with additive manufacturing. IOP conference series: materials science and engineering, June 2021

Materialize (2020) Sourcing low-criticality parts for aircraft. The additive manufacturing opportunity for aerospace supply chain management. https://assets-eu-01.kc-usercontent.com/8ff24b0e-57a3-0157-62d1-fa4ac9734eb5/3cc5290c-2630-4a7e-a930-eb97c43a4671/Whitepaper-additive-manufacturing-aerospace-supply-chain.pdf

Singamneni S, Lv Y, Hewitt A, Chalk R, Thomas W, Jordison D (2019) Additive manufacturing for the aircraft industry: a review. 8:1. https://doi.org/10.4172/2329-6542.1000214

Thoppil NM, Subbu K (2014) Application of rapid prototyping in the aerospace industry. Rapid Manufacturing Processes, Warangal, India, pp.1–11

Türk D-A, Triebe L, Meboldt M (2016) Combining additive manufacturing with advanced composites for highly integrated robotic structures. https://doi.org/10.1016/j.procir.2016.04.202

Vashishtha V, Makade R, Mehla N (2011) Advancement of rapid prototyping in aerospace industry review. Intl J Eng Sci Technol 3

Zhu J, Zhou H, Wang C, Zhou L, Yuan S, Zhang W (2021) A review of topology optimization for additive manufacturing: status and challenges. Chinese J Aeronaut. https://www.sciencedirect.com/science/article/pii/S1000936120304520 https://doi.org/10.1016/j.cja.2020.09.020

Zijm WHM, Knofius N, van der Heijden M (2019) Additive manufacturing and its impact on the supply chain. https://doi.org/10.1007/978-3-319-92447-2_23

Comparison of the Speed Change and Vector Maneuver Techniques for the Conflict Resolution Problem: Fuel and Flight Time Analysis

Kadir Dönmez and Ramazan Kursat Cecen

Nomenclature

APC	Aircraft Performance Category
ATCo	Air Traffic Controller
CRP	Conflict Resolution Problem
FC	Fuel Consumption
FLC	Flight Level Change
NB	Narrow Body
RJ	Regional Jet
SC	Speed Change
TMA	Terminal Maneuvering Area
VM	Vector Maneuvering
WB	Wide Body

1 Introduction

Increasing air traffic demand raises the workload of air traffic controllers (ATCos) (Hah et al. 2006; Yazgan et al. 2021). Considering the current air traffic density, ATCos need decision support systems for conflict resolution in airspace and Terminal Maneuvering Area (TMA) to manage air traffic safely (Dudoit et al. 2022). The

K. Dönmez (✉)
Samsun University, Samsun, Turkey
e-mail: kadir.donmez@samsun.edu.tr

R. K. Cecen
Eskisehir Osmangazi University, Eskişehir, Turkey
e-mail: ramazankursat.cecen@ogu.edu.tr

T. H. Karakoc et al. (eds.), *Emerging Trends in Electric Aviation*, Sustainable Aviation, https://doi.org/10.1007/978-3-031-37299-5_10

safe separation between aircraft pairs is achieved in three different ways that are speed change (SC), vector maneuver (VM), and flight level change (FLC). By using these methods, minimum necessary separations between aircraft are achieved, and conflict resolution is performed (Cecen et al. 2020). FLC is easy to implement but increases the ATCos workload since observing the interaction of traffic at different levels is essential. On the other hand, although the vertical geometry is preserved by using the VM, the aircraft's deviation from their horizontal routes leads to more monitoring and instructions. In addition, this situation also leads to an increase in frequency occupancy, thus increasing the controller workload. Therefore, SC is an advantageous approach in terms of controller workload compared to the methods mentioned above, thanks to the preservation of both horizontal and vertical geometry. Furthermore, studies indicate that this method provides fuel efficiency compared to other methods (Cecen and Cetek 2020).

1.1 Literature Review

In the literature, the conflict resolution problem (CRP) has been handled using linear programming (Niedringhaus 1995), mixed integer programming (Cecen 2021; Dias et al. 2022; Xiangmin et al. 2020; Pallottino et al. 2002), non-linear programming (Alonso-Ayuso et al. 2016b; Cafieri and Rey 2017; Cecen and Cetek 2019), etc. In these approaches, optimum results have been achieved for a smaller number of aircraft. However, in the case of higher aircraft numbers, CRP has been addressed using heuristic algorithms at the expense of finding near-optimal results (Cecen et al. 2020; Durand and Alliot 2009). Deviations from existing routes by resolving the conflicts can cause severe fuel and delay increases for aircraft. Therefore, in the presented studies, minimization of many objectives such as fuel (Vela et al. 2009b), delay (Omer 2015), deviation from the existing route (Hernández-Romero et al. 2017), the total cost of deviations (Vela et al. 2009a), acceleration rate (Alonso-Ayuso et al. 2016b), etc. have been discussed. While some studies handled the conflict resolution methods separately (Pallottino et al. 2002), some make comparisons or consider them together (Alonso-Ayuso et al. 2016a; Christodoulou and Costoulakis 2004). Models in which more than one conflict resolution technique is integrated yielded more flexible and efficient results (Cafieri and Omheni 2017). Cecen et al. (2020) applied the VM and FLC maneuvers together for the CRP and stated that this approach offers more promising results than handling the methods separately. Although using these techniques together provides flexibility, it is essential to compare them with each other and reveal their advantages and deficiencies. Some of the above-mentioned studies have shown that SC is more advantageous than other methods in terms of efficiency.

To validate and improve these efforts, the conflict resolution model presented by Cecen (2021) is enhanced, and SC and VM techniques are handled separately in this study. In addition, more advanced speed restrictions, calculated based on aircraft types, are integrated into the model. Linear regression equations were also integrated into the model, and speed and flight level-dependent fuel calculations were performed for each aircraft type.

2 Method

The CRP is critical in terms of providing a safe air traffic flow. This problem can be experienced at every stage of the flight. This study discusses the en-route phase, the longest part of the flight phases. Two types of conflict can occur in the horizontal plane during the en-route phase. These are aircraft conflicts at the intersecting routes and the same route. These two types of conflicts are shown in Figs. 1 and 2, respectively.

This study considers generic airspace, including two routes and five different flight levels (Fig. 3). There is an intersection point for each altitude. Information about the routes, entry and exit points, and flight levels are integrated into the model as parameters. The length of all airways was determined as 100 nautical miles (NM) and the distance to the intersection point as 50 NM.

The safe separation distance between aircraft, $D_{\min}$, is assumed to be 5 NM. The time separation method is used to maintain safe separation distances between aircraft. Conflicts on the same route are checked at both the entry and exit points. With this method, aircraft are prevented from over-crossing. The safe separation time T_{ij} between two aircraft on intersecting routes varies according to the angle of conflict between the aircraft, θ, and the speeds of the consecutive aircraft, V_i, V_j.

$$T_{ij} = \frac{D_{\min}}{V_i V_j |\sin \theta_{ij}|} \sqrt{(V_i)^2 + (V_j)^2 - 2V_i V_j \cos(\theta_{ij})} \tag{1}$$

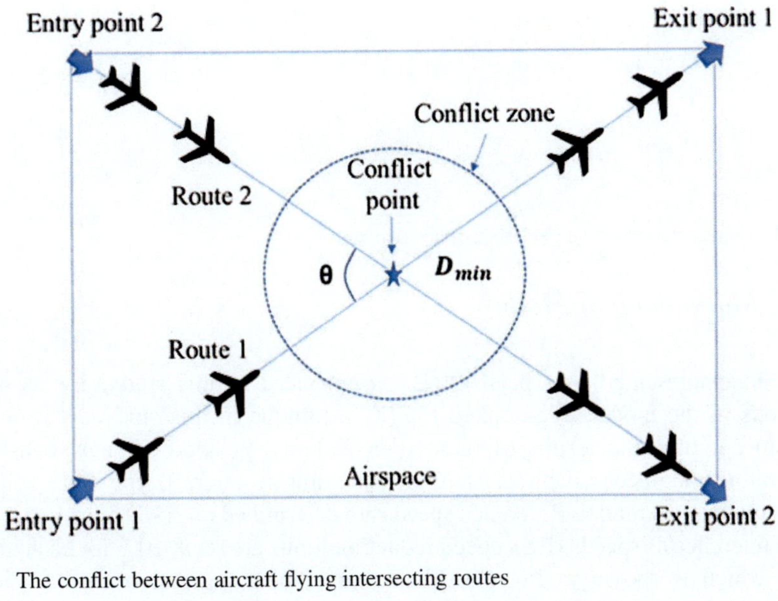

Fig. 1 The conflict between aircraft flying intersecting routes

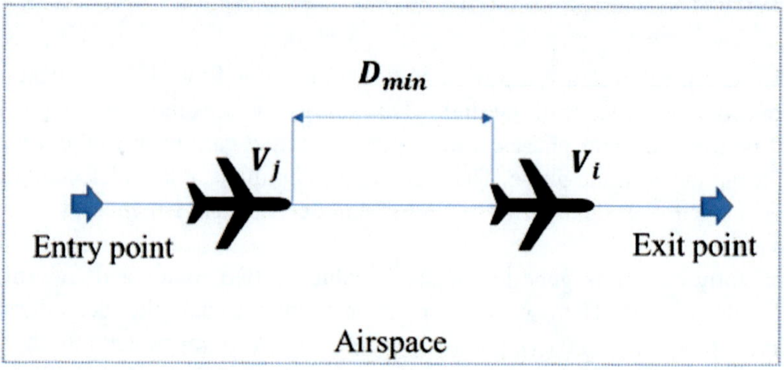

Fig. 2 The conflict between aircraft flying the same route

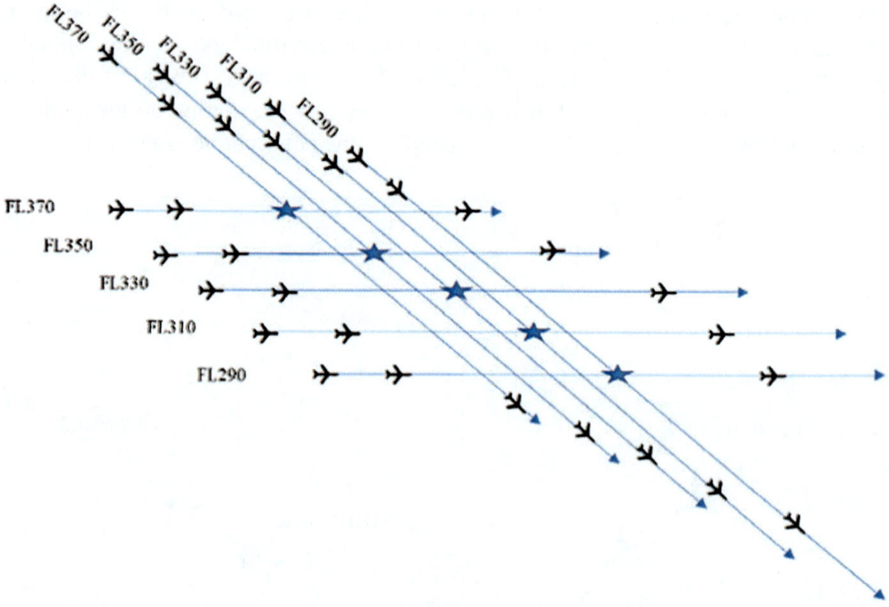

Fig. 3 Representation of the generic airspace structure

2.1 Mathematical Model

The model presented by Cecen (2021) is enhanced in this study. He used VM methods in the model. We applied the SC technique in this study, as well as to VM to compare the results of these two methods. In addition, new constraints controlling the speed change of aircraft according to aircraft type were added to the mathematical model. Reference speeds are determined on BADA 3.11. Based on these reference airspeeds (RS), speed reduction limits are set at 10% for each aircraft type, which is shown in Eq. (2). In addition, linear regression Eqs. (3–5) that calculate fuel consumptions depending on the speed and altitude of each aircraft performance category (APC) are integrated into the model. This study implements three different APCs: regional jet (RJ), narrow-body (NB), and wide-body (WB) aircraft. Speed constraints and fuel calculation equations are given further. To reach the whole model, please see Cecen (2021).

$$0.9 \cdot RS \leq V \leq RS \tag{2}$$

$$p_{00} + p_{01} \cdot A + p_{10} \cdot B \tag{3}$$

$$r_{00} + r_{01} \cdot A + r_{10} \cdot B \tag{4}$$

$$s_{00} + s_{01} \cdot A + s_{10} \cdot B \tag{5}$$

The RS values are determined as 450 knots, 430 knots, and 480 knots for narrow-body aircraft, regional jets, and wide-body aircraft, respectively, where A is the aircraft flight level and B is the aircraft airspeed. p_{00}, p_{01}, p_{10}, r_{00}, r_{01}, r_{10}, s_{00}, s_{01}, and s_{10} are the regression coefficients.

2.2 *Operational Concept*

In this study, it is assumed that the VM is received before entering the airspace. Therefore, the aircraft is delayed before entering the airspace. On the other hand, the SC is applied after the aircraft enters the airspace. Figure 4 reflects an example of conflict resolution methods implemented in the model.

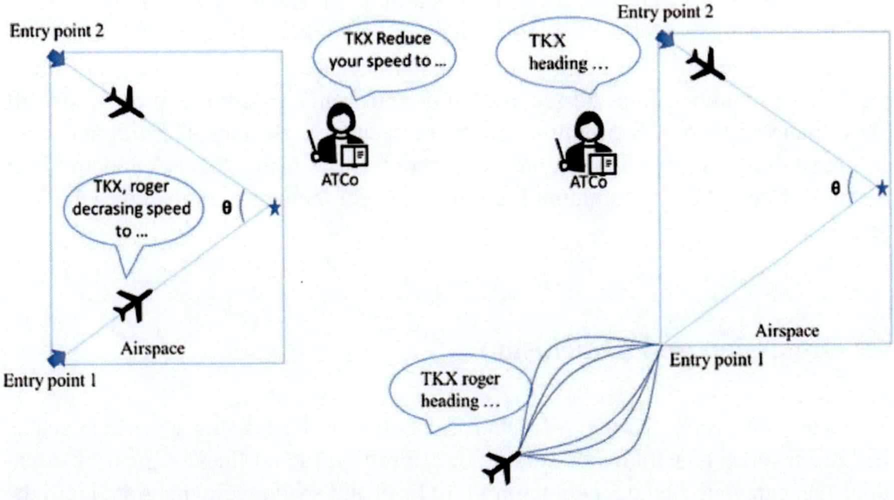

Fig. 4 Operational concept

Table 1 The results of FC for SC and VM approaches

Scenario	SC	SC Ins.	VM	VM Ins.	Savings (%)
1	838.44	40	851.71	31	1.56
2	802.03	50	812.42	21	1.28
3	849.57	49	866.45	32	1.95
4	946.70	39	965.11	40	1.91
5	915.03	44	947.73	50	3.45
6	870.48	46	884.91	30	1.63
7	789.07	41	803.75	36	1.83
8	888.43	47	904.65	31	1.79
9	892.11	39	905.71	34	1.50
10	882.19	31	895.89	29	1.53
Average	867.41	42.6	883.83	33.4	1.84

3 Results

This study ran ten air traffic scenarios for each SC and VM technique. Table 1 shows the results of the mathematical model. The first column shows the scenario number; the second and third columns represent average fuel consumption (FC) per aircraft (kg) using SC and the number of aircraft that used SC. Similarly, the fourth and fifth columns show FC using VM and the number of aircraft using VM methods. Finally, the last column displays the fuel-saving percentage.

According to the result, it was observed that approximately 1.84% fuel saving was achieved by using the SC method compared to the VM. The maximum fuel save was obtained in scenario five since it has the maximum VM instruction number. As it is seen that as the number of VM increases, the fuel consumption rises. In addition, these two methods were compared in terms of flight times. Accordingly, in each sample, an average of 4% flight time increase was observed for the SC technique compared to the VM. The results are presented in Table 2. The first column shows the scenario number, and the second, third, and fourth columns represent aircraft performance category distribution among the scenarios. Besides, the fifth and sixth columns show average flight time per aircraft (sec.) using the SC and the VM methods. Finally, the last column presents the flight increases compared to the VM approach.

4 Discussion and Conclusion

This study compares SC and VM approaches for the CRP using generic airspace. Enhanced speed constraints depending on aircraft types and linear regression equations that calculate fuel depending on flight level and speed were integrated into the model developed. As a result, it is observed that the SC method saves 16.4 kg of fuel

Table 2 The results of flight time for SC and VM approaches

Scenario	NB	RJ	WB	Speed change	Vector maneuvers	FT increase (%)
1	%37	%34	%29	827.44	798.08	3.68
2	%31	%42	%27	847.89	802.04	5.72
3	%40	%30	%30	830.01	796.10	4.26
4	%31	%31	%38	818.36	792.47	3.27
5	%29	%36	%35	828.02	795.82	4.05
6	%30	%38	%32	835.69	798.06	4.72
7	%29	%45	%26	840.65	803.65	4.60
8	%28	%38	%34	833.91	797.06	4.62
9	%38	%29	%33	818.16	794.23	3.01
10	%31	%36	%33	820.02	796.82	2.91
Average	%32	%36	%32	830.02	797.43	4.08

per aircraft at an increase of approximately half a minute in average flight time. In the VM method presented in this study, the bank angles of the aircraft are not considered. Therefore, the technique may lead to more fuel consumption when the bank angle is added to the model.

Consequently, it can be said that the speed change method is more likely to save fuel than the vector technique. Binary combinations of these approaches will also be discussed by adding FLC, SC, and VM to the developed model in future studies. In addition, it is planned to develop a stochastic model by integrating an essential source of uncertainty in ATM, such as wind, into the model in the future.

References

Alonso-Ayuso A, Escudero LF, Martín-Campo FJ (2016a) An exact multi-objective mixed integer nonlinear optimization approach for aircraft conflict resolution. TOP 24(2):381–408. https://doi.org/10.1007/s11750-015-0402-z

Alonso-Ayuso A, Escudero LF, Martín-Campo FJ (2016b) Exact and approximate solving of the aircraft collision resolution problem via turn changes. Transp Sci 50(1):263–274. https://doi.org/10.1287/trsc.2014.0557

Cafieri S, Omheni R (2017) Mixed-integer nonlinear programming for aircraft conflict avoidance by sequentially applying velocity and heading angle changes. Eur J Oper Res 260(1):283–290. https://doi.org/10.1016/j.ejor.2016.12.010

Cafieri S, Rey D (2017) Maximizing the number of conflict-free aircraft using mixed-integer nonlinear programming. Comput Oper Res 80:147–158. https://doi.org/10.1016/j.cor.2016.12.002

Cecen RK (2021) A mixed integer linear programming approach for aircraft conflict detection and resolution problem. Int J Eng Res Dev 13(2):350–358. https://doi.org/10.29137/umagd.736065

Cecen RK, Cetek C (2019) A two-step approach for airborne delay minimization using Pretactical conflict resolution in free-route airspace. J Adv Transp 2019:1. https://doi.org/10.1155/2019/4805613

Cecen RK, Cetek C (2020) Conflict-free en-route operations with horizontal resolution manoeuvers using a heuristic algorithm. Aeronaut J 124(1275):767–785. https://doi.org/10.1017/aer.2020.5

Cecen RK, Saraç T, Cetek C (2020) Meta-heuristic algorithm for aircraft pre-tactical conflict resolution with altitude and heading angle change maneuvers. TOP 29:629. https://doi.org/10.1007/s11750-020-00585-8

Christodoulou M, Costoulakis C (2004) Nonlinear mixed integer programming for aircraft collision avoidance in free flight. In: Proceedings of the 12th IEEE mediterranean electrotechnical conference (IEEE Cat. No.04CH37521), vol 1, pp 327–330. https://doi.org/10.1109/MELCON.2004.1346858

Dias FHC, Hijazi H, Rey D (2022) Disjunctive linear separation conditions and mixed-integer formulations for aircraft conflict resolution. Eur J Oper Res 296(2):520–538. https://doi.org/10.1016/j.ejor.2021.03.059

Dudoit A, Rimša V, Bogdevičius M, Skorupski J (2022) Effectiveness of conflict resolution methods in air traffic management. Aerospace 9(2):112. https://doi.org/10.3390/aerospace9020112

Durand N, Alliot JM (2009) Ant Colony optimization for air traffic conflict resolution. In: Proceedings of the 8th USA/Europe air traffic management research and development seminar, ATM, pp 182–187

Hah S, Willems B, Phillips R (2006) The effect of air traffic increase on controller workload. Proc Hum Factors Ergon Soc Ann Meet 50(1):50–54. https://doi.org/10.1177/154193120605000111

Hernández-Romero E, Valenzuela A, Rivas D (2017) Probabilistic aircraft conflict detection and resolution considering wind uncertainty. SESAR Innovation Days

Niedringhaus WP (1995) Stream option manager (SOM): automated integration of aircraft separation, merging, stream management, and other air traffic control functions. IEEE Trans Syst Man Cybern 25(9):1269–1280. https://doi.org/10.1109/21.400505

Omer J (2015) A space-discretized mixed-integer linear model for air-conflict resolution with speed and heading maneuvers. Comput Oper Res 58:75–86. https://doi.org/10.1016/j.cor.2014.12.012

Pallottino L, Feron EM, Bicchi A (2002) Conflict resolution problems for air traffic management systems solved with mixed integer programming. IEEE Trans Intell Transp Syst 3(1):3–11. https://doi.org/10.1109/6979.994791

Vela AE, Salaun E, Solak S, Feron E (2009a) A two-stage stochastic optimization model for air traffic conflict resolution under wind uncertainty. In: IEEE/AIAA 28th digital avionics systems conference, 2.E.5-1-2.E.5-13. https://doi.org/10.1109/DASC.2009.5347531

Vela A, Solak S, Singhose W, Clarke JP (2009b) A mixed integer program for flight-level assignment and speed control for conflict resolution. In: Proceedings of the 48h IEEE Conference on Decision and Control (CDC) Held Jointly with 2009 28th Chinese Control Conference, January, 5219–5226. https://doi.org/10.1109/CDC.2009.5400520

Xiangmin GU, Renli LV, Liu H (2020) Conflict resolution problems solved with linear programming in air traffic management systems. In: 2020 Chinese Control And Decision Conference (CCDC), pp 2464–2468. https://doi.org/10.1109/CCDC49329.2020.9164496

Yazgan E, Sert E, Şimşek D (2021) Overview of studies on the cognitive workload of the air traffic controller. Int J Aviation Sci Technol 2(01):28–36. https://doi.org/10.23890/IJAST.vm02is01.0104

Assessing Battery Characteristics During a Full Discharge in an Electric Aircraft

Brooke E. Wheeler, Isaac M. Silver, Brian A. Kish, Markus Wilde, and Gaspar Andre

Nomenclature

RFT Remaining flight time
SOC Battery state of charge

1 Introduction

Electric propulsion has been at the center of the aerospace industry's effort to move towards sustainability and achieve net-zero emissions. The use of batteries in aircraft propulsion offers potential to cut not only CO_2 emissions as well as emissions of other greenhouse gases such as NOx (Neuman 2016). Furthermore, battery-powered propulsion enables high energy efficiency and a quieter flight, decreasing environmental impacts of noise.

While the battery technology and management have developed to such an extent that it has become an established niche for all-electric cars, the batteries to power aircraft have not yet reached an ideal landmark, especially with respect to energy

B. E. Wheeler (✉) · M. Wilde · G. Andre
Florida Institute of Technology, Melbourne, FL, USA
e-mail: bwheeler@fit.edu; gandre2020@my.fit.edu

I. M. Silver
Energy Management Aerospace, Melbourne, FL, USA
e-mail: isaac@energymanagementaero.com

B. A. Kish
University of Michigan, Ann Arbor, USA
e-mail: bakish@umich.edu

T. H. Karakoc et al. (eds.), *Emerging Trends in Electric Aviation*, Sustainable Aviation, https://doi.org/10.1007/978-3-031-37299-5_11

storage and battery life management. The maximum commercial battery energy storage currently available has an energy density of 150–250 Wh/kg; an ideal energy density for these batteries would be at least 500 Wh/kg (Tom et al. 2021). In addition, longer battery life cycles and faster recharge speeds are also necessary for battery-powered aircraft to develop.

There are now certified electric aircraft in Europe, which lends further significance to the understanding of performance of electric aircraft, battery SOC, motor power, and battery discharge. EASA-mandated fuel/charge reserves require at least 10 minutes of reserve battery for electric aircraft flying near a single airport. Any test of a full discharge of batteries in electric vehicles must occur on the ground. Chen et al. (2017) presented discharge curves for lithium cobalt batteries, indicating that they were not linear with a steep decline at lower charges. This, combined with safety, means that practical knowledge of both discharge and performance is lacking at low SOCs and is particularly important for electric aircraft.

The purpose of this study was to fully discharge the batteries of an electric aircraft at minimum cruise power and examine the motor power periodically as well as battery temperature and RFT. Understanding battery discharge and the relationship with performance is essential to our knowledge of how pilots will need to adjust to operating electric aircraft and how best to maintain safe operations.

2 Method

The test article is a single engine electric aircraft certified in Europe as an electric light sport aircraft but is not certified in the USA. The aircraft has two liquid-cooled battery packs and an inverter/motor-controller. The power is provided by an AC axial-flux motor, which drives a three blade, fixed pitch propeller. At 100% (full) charge, the powerplant can generate 69 kW of peak power.

After a typical test flight, this experimental run was conducted to fully discharge both battery packs. Because of mandated fuel/power reserves and safety, electric aircraft battery performance cannot be tested to 0% during flight. However, the engine can be run until the batteries fully discharge on the ramp. The aircraft was tied down with chocks behind the wheels during the test which was conducted on the ramp. The starting battery charge read 64%. A 20 kW minimum cruise power was initiated and used to discharge the batteries while minimizing stress on the batteries. Every 10% decrease in battery SOC, the throttle was increased to maximum power to test the motor power. All data output from the aircraft was collected to examine battery temperature management, flight time remaining, battery SOC, motor power, and the overall discharge curve.

3 Results and Discussion

Regular flight tests have indicated battery temperature, SOC, and motor performance decreasing as expected during the battery discharge from 100% down to 50%. This test used a minimum cruise setting to fully discharge both batteries from 64% SOC to 0%; this was accomplished in 37.9 min.

Both batteries (see Fig. 1) and motor temperatures were monitored until the very end of the test. Figure 1 shows that both batteries had approximately the same temperature curves for both minimum and maximum temperatures. The thermal management system worked well overall, but in the final 7 min, the battery temperature increased, reaching 39 degrees as battery SOC approached 0.

Figure 2 depicts both SOC, which is equivalent in both battery 1 and 2, and the estimated flight time remaining. As expected, both curves decrease over the course of the test run. However, initial estimates of flight time in the first few minutes are unrealistically low, and the RFT estimated during the final 7 min is greater than that suggested by the SOC. The dips in the flight time remaining occur during the engine power tests. SOC and flight time remaining did not agree at various times, creating a confounding decision-making environment.

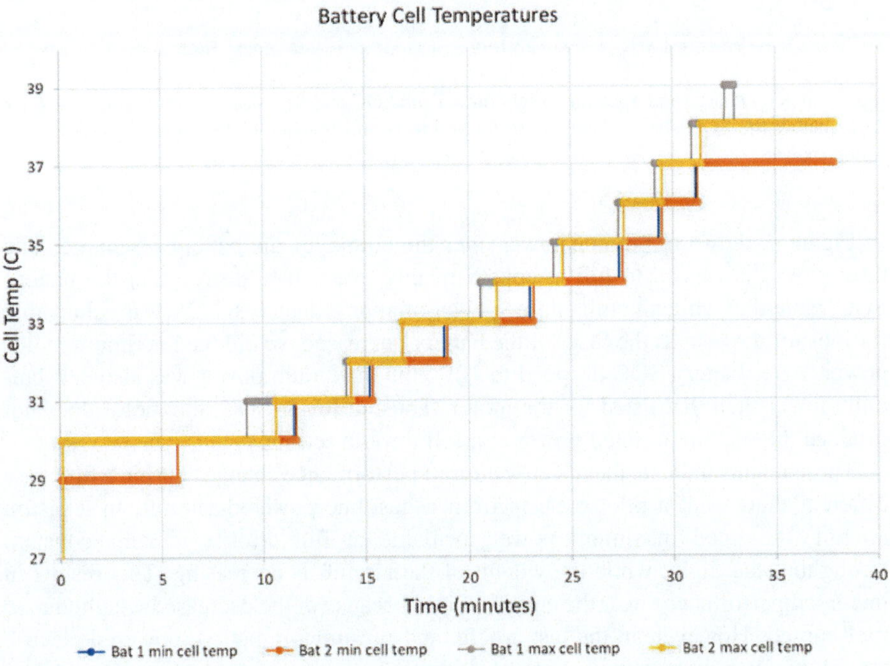

Fig. 1 Battery cell temperatures. Temperatures are well controlled but increasing during the test. Maximum temperatures peak at 39 degrees Celsius when SOC is below 10%

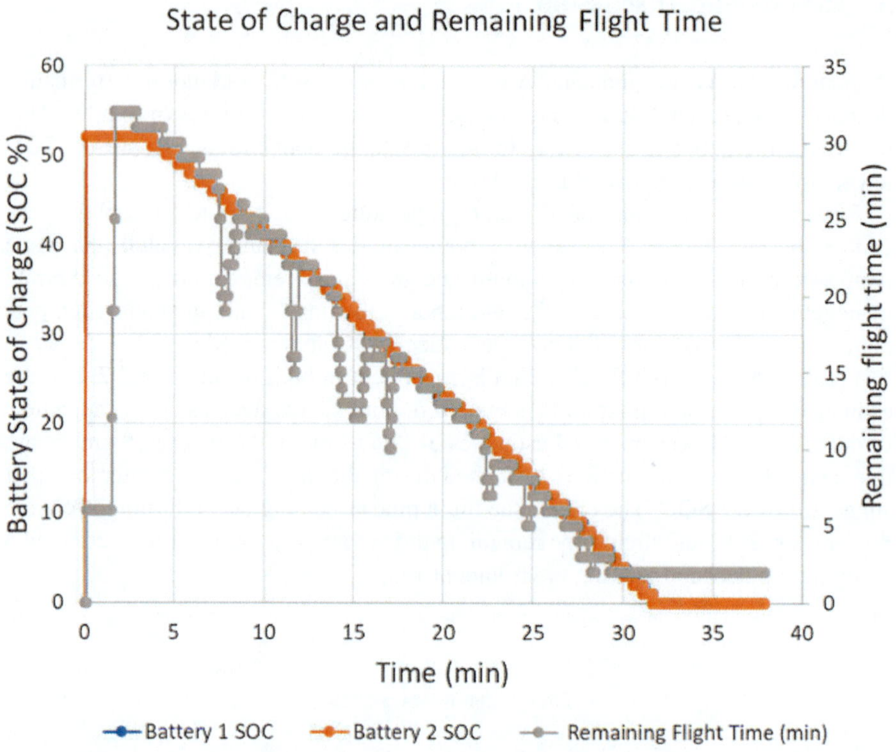

Fig. 2 State of charge and remaining flight time. Both SOC and RFT decrease during the test. RFT drops below the SOC curve at maximum throttle tests and remains higher than expected at low battery power

Figure 3 shows the motor power over the course of the battery discharge. The throttle was advanced to full at approximately every 10% decrease in the battery SOC. Note that while the full engine power of approximately 69 kW was achievable for most of the test, as the SOC of the battery decreased, so did the maximum motor power. After battery SOC dropped to 10%, the maximum power was also substantially lower than that listed for the motor. Thus, at lower SOC, the motor does not produce the maximum rated power at a full throttle setting.

These results indicate that electric aircraft performance changes during a flight are different than what a pilot expects from a gasoline-powered aircraft. In a piston aircraft, the rated maximum power available at full throttle remains constant throughout the flight, while the weight of the aircraft is decreasing. This results in increased performance near the end of a flight because of the decreased weight due to fuel burned. However, in the test article, and presumably battery-powered electric aircraft, the weight remained constant, and there was lower performance (i.e., motor power) at the lower SOC near the end of the test. The longer the aircraft flies at a constant airspeed and throttle setting, the faster the battery discharge; the discharge curve is nonlinear. Thus, a piston-trained pilot's expectation for an aircraft's performance later in a flight will not match an electric aircraft.

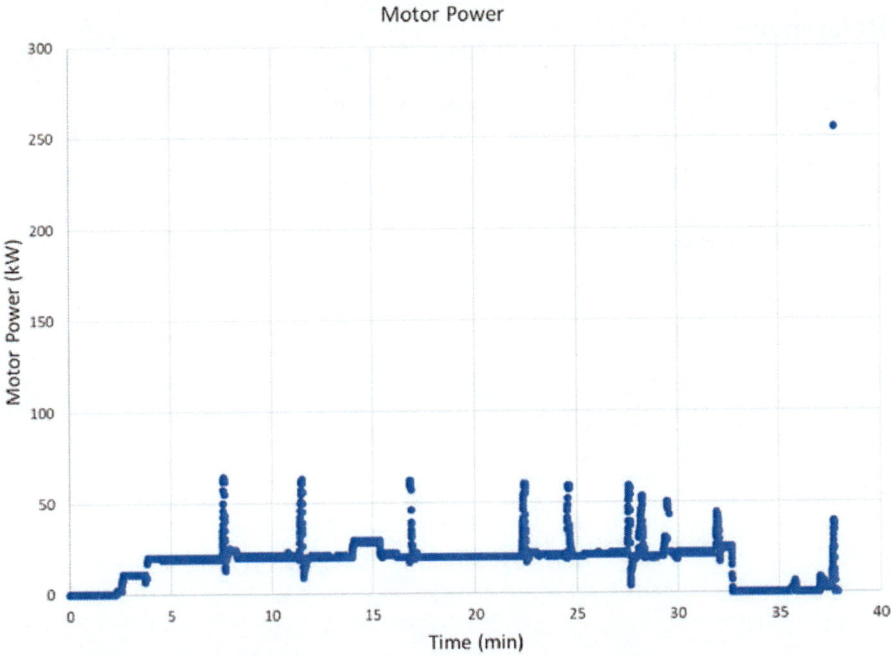

Fig. 3 Motor power. Peak engine power was achievable until lower SOCs

4 Conclusion

There are several important practical implications for electric aircraft. First, as pilots are gaining experience in an electric aircraft, they may be surprised to find that there is not as much usable energy (fuel) remaining as expected at lower battery SOCs. Additionally, there may not be sufficient engine power at full throttle to meet demand in critical maneuvers, such as a go around in the event of an aborted take-off below a certain SOC. The decreased maximum engine power and higher battery temperatures as the batteries' charge approached zero could lead to unexpected system performance. This could also induce mechanical performance anxiety as the performance is not constant across battery charge. Overall, these point to the importance of clear information displays in electric aircraft to aid in decision making, as well as the need for further study of battery discharge and performance, particularly at lower SOCs near the mandated minimum charge reserves.

Acknowledgments This work was supported in part by FAA Contract 692 M15-21-T-00039 Electric Aircraft Trajectory Flight Test Data.

References

Chen Y, Macii E, Poncino M (2017) Workload-driven frequency-aware battery sizing. In: 2017 IEEE/ACM International Symposium on Low Power Electronics and Design (ISLPED), pp 1–6. IEEE

Neuman T (2016) Fly the electric skies. IEEE Spectr 53(6):44–48

Tom L, Khowja M, Vakil G, Gerada C (2021) Commercial aircraft electrification—current state and future scope. Energies 14(2021):8381. https://doi.org/10.3390/en14248381

The Autonomous Air-Sea-Interface-Vehicle: Is It the Key to Abundant Green Energy?

Max F. Platzer and Nesrin Sarigul-Klijn

1 Introduction

Last year, at the United Nations COP 26 Conference in Glasgow, UN Secretary General Antonio Guterres delivered the following statement:

"Even if the recent pledges were clear and credible – and there are serious questions about some of them –we are still careening toward climate catastrophe. Our planet is talking to us. We must listen, and we must act."

Figure 1 is one example how the planet is talking to us. The explosive carbon dioxide increase is an ominous warning signal that should frighten every aerospace engineer. After all, we should never forget the wishful thinking that led to the Challenger and Columbia space shuttle failures in 1986 and 2003 because deviations from the design condition were accepted as the "new normal"!

It is the objective of this symposium to pay attention to Mr. Guterres' admonition and to explore the possibilities of switching to emission-free aircraft propulsion systems, thereby contributing toward the reduction of the harmful carbon dioxide emissions that threaten our survival. However, it is important to keep in mind that the electrical energy or the hydrogen needed to propel electric or hydrogen-powered aircraft needs to be produced from renewable energy sources without emission of greenhouse gases.

In this paper, we hope to interest our aerospace engineering colleagues in the development of a "new" type of vehicle that extracts power from the ocean while flying above the ocean surface. We put the word "new" between quotation marks because such vehicles already exist in the form of sailboats equipped with small hydrokinetic turbines. We are arguing that this small-scale power production

M. F. Platzer (✉) · N. Sarigul-Klijn
Davis, Mechanical and Aerospace Engineering Department, Innovative Power Generation Systems (iPGS) Research Lab, University of California, Davis, CA, USA
e-mail: nsarigulklijn@ucdavis.edu

T. H. Karakoc et al. (eds.), *Emerging Trends in Electric Aviation*, Sustainable Aviation, https://doi.org/10.1007/978-3-031-37299-5_12

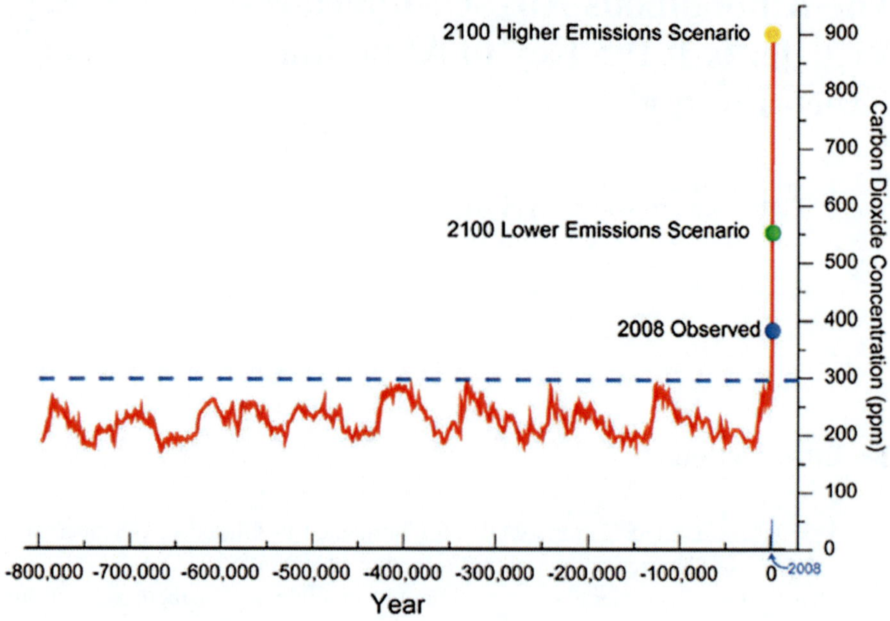

Fig. 1 800,000 Year record of CO_2 Concentration. (Source: GlobalChange.gov)

concept can be greatly expanded to MegaWatt power scales using the latest advances in hydrofoil and sail technology, hydrokinetic turbine technology, and autonomous guidance, navigation, and control technology to develop an "air-sea-interface" vehicle that can expose the turbines to water speeds of 10 or more meters per second.

2 The Energy Conundrum

Two facts are important to understand the conundrum we are facing. In the last hundred years, the global population quadrupled from 2 billion in the 1930s to today's almost 8 billion. This explosive growth was accompanied by an equally explosive growth of the energy available to increase the living conditions. MacKay in his book (MacKay 2009) presented quantitative information about the various amounts of energy needed to maintain the standard of living of the average British citizen. He chose to use the kiloWatthour (kWh) as the energy unit most people can easily relate to. He alerted his readers to the fact that the energy consumption of the average citizen in the developed countries increased from about 20 kWh per person per day in the 1700s to about 100 and even to 300 kWh/p/d in countries such as Canada and Australia since the introduction of the steam engine in the late 1700s. Unfortunately, a large part of the global population is still forced to live at pre-industrial energy supply levels.

Nevertheless, until the end of the last century, there was a justifiable expectation that fossil-based power generation would eventually be able to lift the living standard of most people because the supply of coal, oil, and gas will last for several hundred more years. Warnings of climate scientists about the harmful effect of continued CO_2 emissions were dismissed with the argument that they are exaggerated or, if true, there remains plenty of time for corrective action.

Unfortunately, we now realize that the rate of climate change is much greater than anticipated forcing us to achieve a global transition to emission-free power generation no later than 2050 if irreversible climate change is to be averted. Also, we now realize that the transition to emission-free power generation is very difficult to achieve for technical, socioeconomic and political reasons. These difficulties have been discussed in many papers and books. We merely quote here the engineering professor Frank Incropera of the University of Notre Dame (Incropera 2016): "Reducing emissions will require a sharp departure from the current norm of associating living conditions with levels of consumption. It will also require all hands on deck. Will the effects of global warming be manageable? At this time, there are no answers to these questions. But what can be said is that, through its actions or inactions, humankind will determine its own destiny."

3 Can Aerospace Engineering Come to the Rescue?

Power engineering and aerospace engineering are regarded as two separate disciplines. The former is concerned with the conversion of natural energy resources into usable energy, the latter with the conversion of usable energy into thrust to move an aerospace vehicle. Both disciplines have yielded very impressive results. As aerospace engineers, we often summarize the accomplishments of our discipline as having succeeded in "conquering space and time" by virtue of having shown that hundreds of people can be transported over thousands of miles in less than a day and anyone on this globe can communicate with anyone else within seconds. This was accomplished by the development of airplanes and rockets. The question arises whether aerospace engineering can conquer the "energy barrier."

It is our view that aerospace engineers need to start thinking about vehicles that can produce power rather than requiring thrust, while power engineers need to start thinking about power plants that can move rather than being firmly anchored to the ground. A look at Fig. 2. shows the ocean areas that are exposed to strong and persistent winds. This wind power exceeds the needed global power by a factor of ten. Therefore, tapping into this huge wind energy reservoir offers the opportunity to satisfy the global energy needs for the indefinite future. A further critical advantage is the fact that the oceans belong to no particular nation. On the contrary, the Law of the Seas guarantees all nations equal access to this energy resource. As a consequence, the inherent inequalities between nations that, by geographic accident, own or lack fossil-based resources are eliminated. The exploitation of ocean wind energy therefore will end the competition for nationally owned energy resources, including

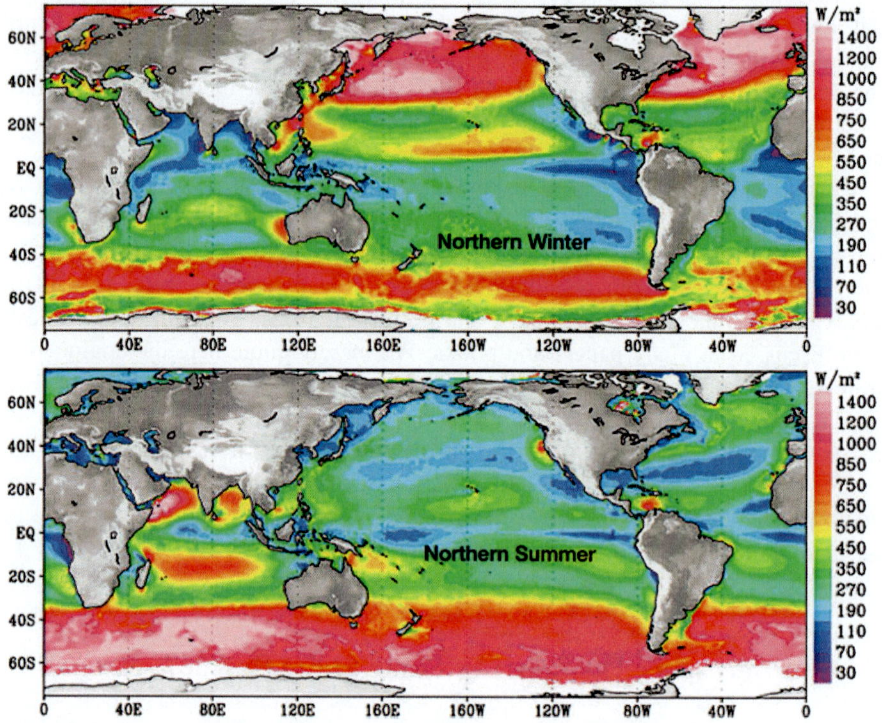

Fig. 2 Ocean wind power distribution. (Source: NASA Source Observatory)

the temptation to use possession of such resources as economic weapons, as is currently happening.

Given this opportunity to "democratize" the world's energy supply, the critical question is how to tap into the ocean wind power. Understandably, power engineers think in terms of ever increasing wind turbines that are firmly anchored to the ground on land. The disadvantages are well appreciated by now. Wind over land is significantly less intense than wind over water, and the wind turbine noise output severely limits their placement in populated areas. For these reasons, their placement into shallow off-shore waters is becoming increasingly attractive, especially in Europe. The logical next step is to capture the wind by floating the turbines instead of anchoring them to the sea floor. Quite apart from the technical difficulties involved in erecting and stabilizing the huge turbines, the capacity factor of floating wind turbines is limited by the wind conditions at a given fixed location. Furthermore, floating wind turbines far off-shore have to survive occasional heavy storms, further complicating the stabilization challenge that already exists for floating turbines closer to shore.

The question therefore arises whether the direct conversion of wind into electric power by wind turbines can be replaced by a device that uses both wind and water in the same system. In such a system, the wind is used to propel the device to as high a speed in water as possible and to mount hydrokinetic turbines for the generation of electric power. The first part has been implemented for millennia in various sailing ship configurations, whereas the second part has been demonstrated as yet only at a very small scale. However, it is clear that its extension to larger scales presents no technical problem if ways can be found to store the generated electricity on the sailing ships. In our view, the most obvious solution is the splitting of the sea water into "green hydrogen" and to store it in compressed or liquefied form for eventual transfer to shore.

There is no question that this type of renewable power generation system is more complex and expensive than a typical wind turbine system. However, it is important to remember that it delivers storable energy, and therefore, the comparison needs to be based on comparing it with a wind turbine system in combination with a suitable energy storage system. Furthermore, the comparison needs to be based on the capacity factors of the two systems. Land- and offshore-based wind turbines are likely to have much lower capacity factors than sailing ship-based power generation systems because the latter can operate in high wind areas for longer periods of time. Also, the operation of sailing ships in international waters is only subject to the law-of-the-sea regulations, whereas land- and offshore-based wind turbines require a lengthy approval process. We have described the technical, logistical, and economic aspects of producing storable hydrogen using autonomously operating "energy ships" (as we call them) in some detail in a recent book (Platzer and Sarigul-Klijn 2021). We merely want to point out here that energy ships have a decisive advantage over pure wind energy systems by being able to use the much higher density of water. Furthermore, we draw attention to the critical importance of the ship speed. A doubling of the ship speed translates into an eight times larger power output. For this reason, we prefer to use the word "air-sea-interface-vehicle" instead of sailing ship to draw attention to the need for maximizing the vehicle speed and to minimize its hydrodynamic drag. Hydrofoil boats, such as the ULTIM trimaran boats (Ultim 2022), have already demonstrated their ability to sail at speeds greater than 20 m/s in the open ocean areas exposed to the high wind speeds shown in Fig. 2.

These developments clearly indicate that all the technology elements are available to build new types of "air-sea-interface-vehicles" at sizes and in quantities that enable the transition to a hydrogen-based economy. It is a challenge similar to the moon landing challenge, but the stakes are much higher. We appeal to our aerospace engineering colleagues and students to take up this challenge.

References

Incropera FP (2016) Climate change: a wicked problem. Cambridge University Press. https://doi.org/10.1017/CBO9781316266274

MacKay, DJC (2009) Sustainable energy – without the hot air. UIT Cambridge http://www.withouthotair.com/cft.pdf

Platzer, MF, Sarigul-Klijn, N. (2021) The green energy ship – renewable energy from wind over water. Springer Briefs in Applied Sciences and Technology https://link.springer.com/content/pdf/10.1007%2F978-3-030-58244-9.pd

ULTIM Racing Trimarans (2022). https://www.boatsnews.com

Development of Viscous CFD Analysis Model Including Real Gas Effects for Nose Optimization at Hypersonic Speeds

Ali Alperen Özkan and Atilla Bıyıkoğlu

1 Introduction

In missile design, it is important to create an ideal nose geometry that reduces the aerodynamic heating in the nose area and the wave drag force. CFD plays a crucial role while creating the ideal nose geometry especially at hypersonic speeds. The studies in the hypersonic fields mostly cover developing schemes since obtaining proper solution is difficult in hypersonic region. To illustrate, Liou and Steffe (Liou and Steffen 1993) developed the Advection Upstream Splitting Method (AUSM) discretization scheme that may be suitable for gases satisfying the equation of state that do not contain a Jacobian matrix. Liou (1996) realized the shortcomings of the AUSM and developed a flux scheme called AUSM+. Dala (1997) examined hypersonic viscous flows, which include non-equilibrium real gas effects. Ismail et al. (2009) proposed a flux scheme based on the conservation of entropy to solve the shock release problem. Darwish and Moukalled (2014) developed and tested an algorithm to solve fluid problems at all speeds. Pişkin and Eyi (Piskin and Eyi 2015) studied chemical, thermal, and vibrational imbalance modeling. Each mode was analyzed with the assumptions of one temperature and two temperatures. Chen et al. (2018) developed a flux discretization scheme called AUPM based on the Zha-Bilgen approach to overcome the problem of hypersonic heating predictions.

Examination of the studies in the hypersonic fields shows that developing CFD analysis model for nose optimization was not carried out in detail. To fill the gap in the literature, this study aims to develop a CFD analysis model to be used in the nose optimization at hypersonic speeds and to validate the model.

A. A. Özkan (✉) · A. Bıyıkoğlu
Department of Mechanical Engineering, Gazi University, Ankara, Türkiye
e-mail: alialperen.ozkan@gazi.edu.tr; abiyik@gazi.edu.tr

T. H. Karakoc et al. (eds.), *Emerging Trends in Electric Aviation*, Sustainable Aviation, https://doi.org/10.1007/978-3-031-37299-5_13

2 Analysis Model

The geometry used during the verification was Standard Hypervelocity Ballistic-1 (HB-1). Validation geometry was modeled parametrically based on its diameter in the Siemens NX environment. The dimensions of the created model and the 3-D view of the model are given in Fig. 1.

Previous field experience and similar studies were considered while creating the solution grid for HB-1. It was aimed to obtain a grid that is suitable for the "low level" CFD analyses. Required features used in grid generation process are summarized in Table 1. For the advanced CFD analysis, the appropriate grids should be obtained after grid independency study.

Generated grids for the CFD analyses are given in Fig. 2. Left side of Fig. 2 details O-grid structure that is used to eliminate skewness problem at base and nose part. In Fig. 3, close view of boundary layer structure can be seen.

3 Methodology

Methodology of the study includes numerical solver, governing equations used in the solver, properties of fluids, boundary conditions, and turbulence modeling and simulation.

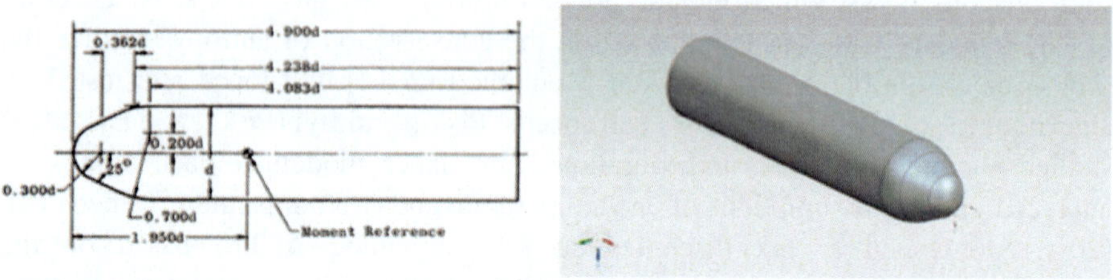

Fig. 1 Standard Hypervelocity Ballistic-1 Correlation Configuration (left: dimensions, right: 3D-modelling)

Table 1 Main considerations during grid generation

Geometric component	Required features
Reference length	Body diameter or base diameter of nose section for missiles
Blunt nose	Element size: 3–5% of nose diameter, constant
Base	Element size: 5–10% of reference length
Body	Element size: 15–20% of reference length
External domain	Shape: Cylinder, length: 20 times the model length, radius: 7 times the model length

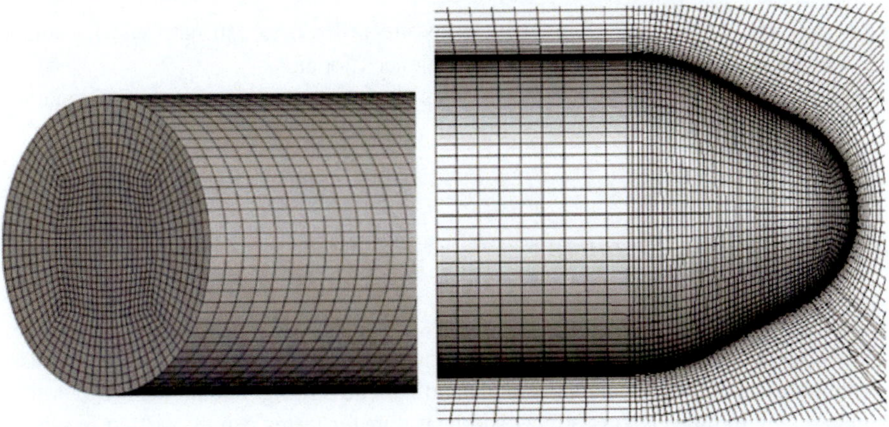

Fig. 2 Base O-grid structure(left), body and nose grids (right)

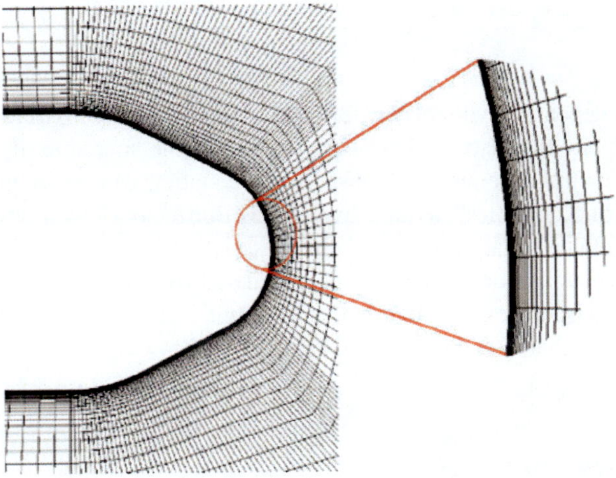

Fig. 3 Close view of boundary layer structure (First layer thickness = 1e-06, Growth ratio = 1.2)

3.1 Numerical Solver

In the study, METACOMP CFD++ v20.0 is used. According to Chakravarthy (Chakravarthy 1999) and Hirsch (2007), capabilities and limitations of the solver are:

- Steady and/or unsteady compressible and incompressible Navier–Stokes equations with various turbulencemodels,
- Total variation diminishing (TVD) discretization that gives oscillation-free solutions in CFD,
- Riemann solvers for better signal propagation in flow field,

- Non-TVD derivatives of inviscid terms are utilized to estimate viscous terms accurately and discretization of theturbulence model,
- Variety of pointwise turbulence models that do not depend upon information about y+ values.

3.2 Governing Equations

Base equation used in the study is compressible real gas Navier-Stokes equations. For Newtonian fluid satisfying the continuum condition, conservation of mass, momentum, and energy equations are called "governing equations." Governing equations including viscous effects and heat transfer terms can be written in closed form as in Eq. 1

$$\frac{\partial u}{\partial t}+\frac{\partial (E_i-E_v)}{\partial_x}+\frac{\partial (F_i-F_v)}{\partial_y}+\frac{\partial \left(G_i-G_y\right)}{\partial_z}=\dot{\mathrm{S}} \tag{1}$$

Where *U* is dependent variable vector; *Ei*, F*i*, and *Gi* are the inviscid flux vectors; *Ev*, F*v*, and *Gv* are the viscous flux vectors; and *S* is the source term. The detailed information about components of conservation equation can be found in "Hypersonic Aerothermodynamics" written by John J. Bertin in 1994. For a Newtonian fluid, there is a linear relationship between stresses and strains. Heat fluxes in three-dimensional Cartesian coordinates can be calculated from Fourier's law of conduction. Dynamic viscosities and thermal conductivities can be calculated by using Sutherland's Law that is embedded in CFD++.

3.3 Properties of Fluids

In hypersonic region, kinetic energy in flow field tends to increase with increasing velocity. With increasing energy, density is no longer a function of temperature and pressure only. Under this circumstances, real gas assumption is used to define the equation of state since ideal gas behavior fails.

Instead of ideal gas state, cubic equation of state is used for modeling real gases. In the study, van der Waals equation state, which can be calculated from Eq. 2, is used. The values of *a*, *b* as well as the critical compressibility factor *zc* can be obtained from the conditions at the critical point, where the first and second derivatives of the pressure with respect to the specific volume become zero. The values of the constants for the model are given in Table 2.

Table 2 Generalized parameters for van der Waals model using critical properties (Poling et al. 2001)

Equation	$\nu\tilde{}$	$\tilde{}$	z_c	$a(RT_c)^2/2_{pc}$	bRT_{cp}
Van der Waals	0	0	0.375	0.42188	0.125

Table 3 Boundary conditions of validation case

Boundary	Boundary condition	Thermal condition	Turbulence condition
Open domain	Characteristics-based inflow/ outflow M = 8.09 Re_D = 1.1e+07/m	$T_\infty = 54.68$ K	Turbulence/laminar viscosity ratio = 5
Nose	Viscous (no-slip) wall function	Adiabatic	-
Body	Viscous (no-slip) wall function	Adiabatic	-
Base	Viscous (no-slip) wall function	Adiabatic	-

$$p = \frac{RT}{\tilde{v} - b} - \frac{\alpha}{\tilde{v}^2 + 2b\tilde{v} + \tilde{\omega} b^2} \tag{2}$$

3.4 Boundary Conditions

Available boundary conditions in METACOMP CFD++ can be classified in five groups: inflow and outflow, only outflow, wall, symmetry, and zonal. The boundary conditions that are used in the study can be found in Table 3.

3.5 Turbulence Modeling and Numerical Simulation

CFD analyses are run by using realizable k-ε turbulence model where k is kinetic energy and ε stands for dissipation rate. Since the flow is hypersonic, non-equilibrium wall function type is used. To realize diffusive mixing in caused by high inertia, which is nature of hypersonic flow, wall-bounded compressibility correction is activated during analyses.

Steady-state implicit simulation strategy is chosen for all cases and 7500 iterations are done. During analyses, Courant number is ramped from 0.1 to 10 with an adjustment factor of 0.95. Since nature of hypersonic flows causes strong shocks, discretization is done by first order up to some iteration, and then it is blended into second order fully.

3.6 Grid Independency

Estimating and decreasing order of errors and obtaining suitable convergence rate are some key points in CFD field. As mentioned in Sect. 2, optimal grid should be generated for advance CFD analysis. For grid independency studies, four different grids were analyzed at 5° angle of attack, and boundary conditions for analyses are given in Table 3.

Roache (1997) suggested a method by using Richardson's extrapolation called grid convergence index, and it is calculated from Eq. 5. For grid independency, C_m is used to calculate ε. Factor of safety, Fs, is taken 1.25 since there is more than 3 grids (Hassanien et al. 2017).

$$\varepsilon = \frac{C_{m,coarse} - C_{m,fine}}{C_{m,fine}} \tag{3}$$

$$p = \frac{\ln \left|(C_{m,coarse} - C_{m,medium})/(C_{m,medium} - C_{m,fine})\right|}{\ln r} \tag{4}$$

$$\mathrm{GCI} = F_S \frac{\varepsilon}{r^p - 1} \tag{5}$$

Variation of pitching moment coefficient, C_m, with angle of for different cell counts is given at Fig. 4. Tables 4 and 5 summarize grid independency study. Higher GCI value means larger error in grid. As seen below, G_{2-3} has lower GCI than G_{1-2}, that is, there is no need to produce finer mesh. Since C_m and wall y^+ values remain constant with increasing grid sizes, fine grid is chosen for further analyses. Also, increasing grid count causes overshooting problems in estimating aerodynamic coefficients.

4 Results and Discussion

After selected fine grid, CFD analysis was run according to conditions given in Table 3. Angle of attack was swept from 0° to 9° with 1 degree interval. In all analyses, side slip angle was taken as zero. Obtained coefficients from CFD analyses were compared and validated with wind tunnel test results that were conducted by Gray and Lindsay (1963) in von Kármán Gas Dynamics Facility. Uncertainties of measurements during wind tunnel tests are given in Table 6.

In Fig. 5, pressure distribution for all grids is given respectively. Results showed that pressure distribution on nose area does not change with the number of elements of grid.

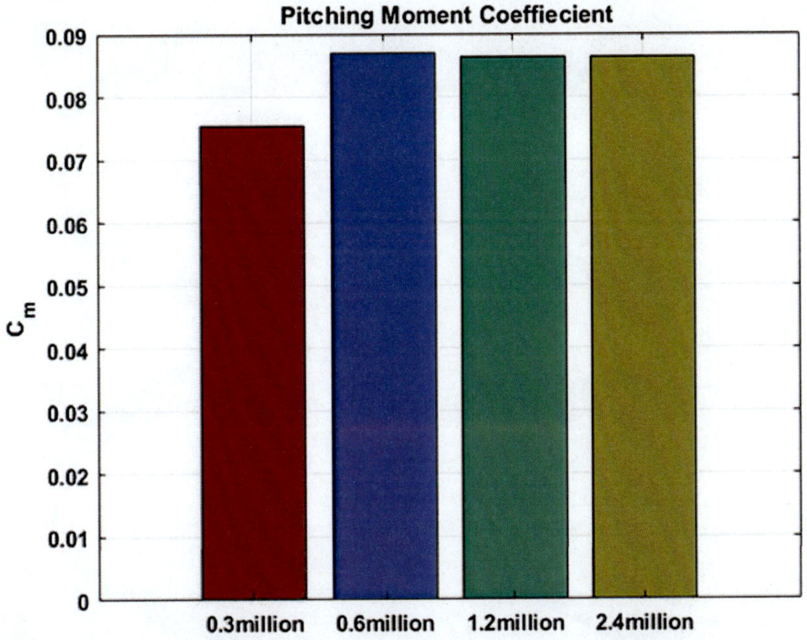

Fig. 4 Change of pitching moment with grid sizes at AoA = 5° and AoSS = 0°

Table 4 Feature of grids for grid independency study

Grids	Cell count	First layer thickness	Refinement ratio (r)	Wall y^+
Coarse	0.3e+06	1e-06	1.2599	0.0649
Medium	0.6e+06	1e-06	1.2599	0.0606
Fine	1.2e+06	1e-06	1.2599	0.0536
Very fine	2.4e+06	1e-06	1.2599	0.0537

Table 5 Grid independency results

Grid direction	p	ε	GCI (%)
G_{1-2}	2.0982	0.028	5.6104
G_{2-3}	16.2731	0.017	0.0505

Table 6 Force capacities of balances and the corresponding statistical uncertainties (Gray and Lindsay 1963)

Balance	5	6
Normal force F_N [lb]	100 ± 0.23	300 ± 0.54
Axial force F_A [lb]	30 ± 0.044	100 ± 0.440
Pitching moment M_Y [in.-lb]	400 ± 0.29	1750 ± 1.36

As can be seen from top left of Fig. 6, Comparing CFD results and wind tunnel data for axial force coefficient (C_X) indicated that CFD analyses estimate this coefficient 5% lesser, since the sting effect was not modeled in analyses.

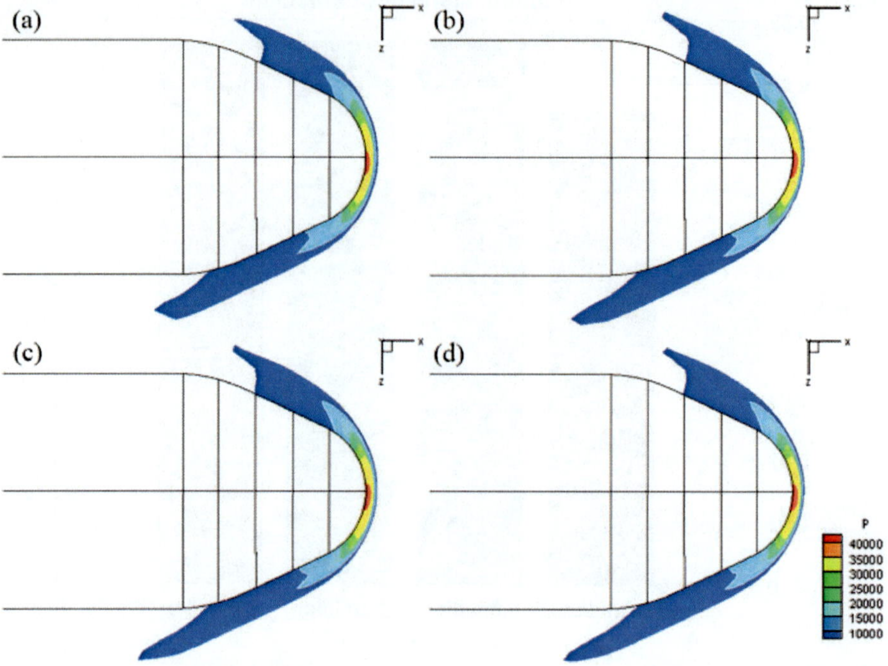

Fig. 5 Pressure distribution at AoA = 5° for different cell counts (**a**) 0.3 million, (**b**) 0.6 million, (**c**) 1.2 million, (**d**) 2.4 million

Examining normal force coefficient (C_Z) and pitching moment coefficient (C_m) showed that the results obtained from CFD analyses had a similar trend to the wind tunnel results.

5 Conclusion

CFD analysis model that can be used in the nose optimization at hypersonic speeds is developed and validated. During analysis, METACOMP CFD++ solver is chosen since it covers hypersonic flows well. Compressible real gas Navier-Stokes equations including viscous effects are used. To validate model, CFD analyses results are compared with wind tunnel results for HB-1. Results show that

- Pressure distribution on nose area does not change with the number of elements of grid.
- CFD analyses estimated C_X 5% lesser since the sting effect was not modeled in analyses.
- C_Z and C_m results obtained from CFD analyses had a similar trend to the wind tunnel results.

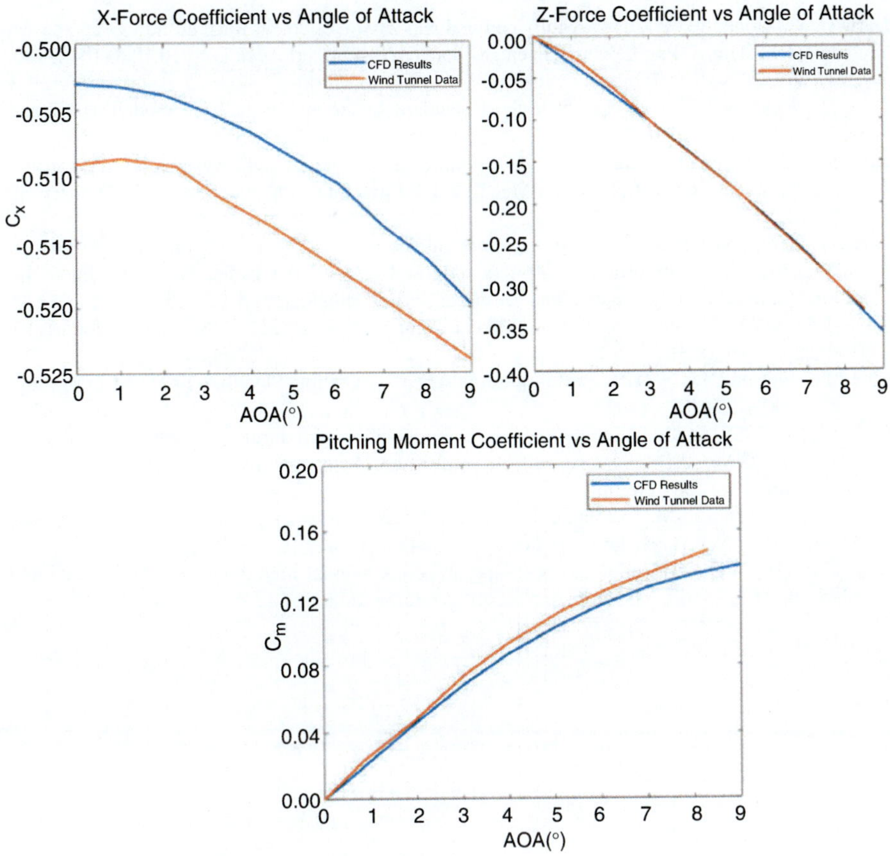

Fig. 6 Variation of x-force coefficient (top-left), z-force coefficient (top-right) and pitching moment coefficient (bottom) with angle of attack

To be concluded, CFD analysis model, which was validated by wind tunnel test data, is suitable for use in future nose optimization studies.

References

Chakravarthy S (1999) A unified-grid finite volume formulation for computational fluid dynamics. Int J Numer Methods Fluids 31(1):309–323

Chen S, Yan C, Zhong K, Xue H, Li E (2018) A novel flux splitting scheme with robustness and low dissipation for hypersonic heating prediction. Int J Heat Mass Transf 127:126–137. https://doi.org/10.1016/j.ijheatmasstransfer.2018.06.121

Dala L (1997) Hypersonic viscous flows including non-equilibrium real gas effects. Doctoral dissertation

Darwish M, Moukalled F (2014) A fully coupled Navier-stokes solver for fluid flow at all speeds. Num Heat Transf Part B: Fundamentals 65(5):410–444. https://doi.org/10.1080/10407790.2013.869102

Gray JD, Lindsay EE (1963) Force tests of standard hypervelocity ballistic models HB-1 and HB-2 at Mach 1.5 to 10. https://apps.dtic.mil/sti/pdfs/AD0412651.pdf

Hassanien IH, Shabaan S, Khalil EE (2017) Numerical simulation and investigation of an atrium fire experiment. Int J Res Chem Metall Civil Eng 4(1). https://doi.org/10.15242/ijrcmce.e0317028

Hirsch C (2007) Numerical computation of internal and external flows, 2nd edn. Elsevier, Oxford

Ismail F, Roe PL, Nishikawa H (2009) A proposed cure to the carbuncle phenomenon. In: Computational fluid dynamics 2006. Springer, Berlin/Heidelberg, pp 149–154

Liou M-S (1996) A Sequel to AUSM: AUSM+. J Comput Phys 129(2):364–382. https://doi.org/10.1006/jcph.1996.0256

Liou M-S, Steffen CJ (1993) A new flux splitting scheme. J Comput Phys 107(1):23–39. https://doi.org/10.1006/jcph.1993.1122

Piskin T, Eyi S (2015) Modelling thermochemical nonequilibrium during atmospheric re-entry. In: 51st AIAA/SAE/ASEE joint propulsion conference. American Institute of Aeronautics and Astronautics, Reston, Virginia

Poling BE, Prausnitz JM, O'connell JP (2001) The properties of gases and liquids, 5th edn. Mcgraw-Hill, New York. https://doi.org/10.1036/0070116822l0

Roache PJ (1997) Quantification of uncertainty in computational fluid dynamics. Annu Rev Fluid Mech 29(1):123–160. https://doi.org/10.1146/annurev.fluid.29.1.123

Real World Path Generation for Non-holonomic Systems with Obstacle Avoidance Using RRT* and Google Earth

S. Sapthagirivasan, M. Seshath, S. Srivarshan, and T. V. K. Sushil Kumar

Nomenclature

GPS Global Positioning System
RRT Rapidly Exploring Random Trees
UAV Unmanned Aerial Vehicles

1 Introduction

Since the Second World War, UAVs have been used for many military operations, which were too risky for piloted missions. Autonomous control of UAVs has been a long-discussed topic because of the non-availability of continuous communicational links, precision maneuvering, and high landing speed (Beard and McLain 2012). Many control systems have been developed for autonomous control of UAVs (Jha 2016; Collinson 2013). These control systems with the help of navigation algorithms can maneuver the UAVs for landing, take-off, or predetermined path flying.

Navigational algorithms are programs that generate a path between the starting point and ending point which are determined by the user. It can also be used to fly around obstacles, which is one of the most important capabilities of autonomous control of UAVs. The obstacles are also predetermined by the user. Presently, A* and RRT are the most widely used algorithms.

S. Sapthagirivasan (✉) · M. Seshath · S. Srivarshan · T. V. K. S. Kumar
Department of Aerospace Engineering, Amrita School of Engineering, Coimbatore, Tamilnadu, India
e-mail: tvk_sushilkumar@cb.amrita.edu

T. H. Karakoc et al. (eds.), *Emerging Trends in Electric Aviation*, Sustainable Aviation, https://doi.org/10.1007/978-3-031-37299-5_14

In this study, GPS and Navigational algorithms were combined to generate a path that had the starting and ending points determined by the user. The path also avoided obstacles.

1.1 Research Gap

Based on the literature survey carried out above, it was found that RRT has been done only in 2D with obstacles and in 3D without obstacles. The method used here was plane line intersect. RRT was not specifically combined with Google Earth to generate a real-word path with obstacle avoidance for non-holonomic systems (Kwak and Sung 2018).

2 Method

The RRT method is as its name says, random points are generated and connected to the closest node that is available. Each time the points are joined, a check must be made so that the line lies outside the obstacle (Lee et al. 2012). Further, all the points that are connected must avoid obstacles. The algorithm stops when a node is near the goal region or when the number of nodes gets over.

RRT* is an optimized version of RRT. When the number of nodes approaches infinity, RRT* will give the shortest possible path to the goal, but it is not feasible in real life (Nasir et al. 2013). A circle of radius r is formed and the center of the circle is considered the new node (vnew). Instead of finding the closest node, RRT* finds all the nodes inside the circle. The shortest path from starting point to vnew was checked from all the nodes within the circle and joined which is taken as a new path. This is repeated until the circle is near the goal region.

In 3D obstacle, avoidance in RRT* is obtained by using the "plane-line intersect" function (Fox et al. 2002; Islam et al. 2012). In this function, the intersection point of the plane and line is found. Further, the plane is limited to a square which is the face of the cube. Similarly, the intersection point is found for the other five faces. The path is considered to be avoiding obstacles only if there is no intersection point for all the faces of the cube, and using this method, a random path is created for the holonomic system in MATLAB (Gan et al. 2021; Vonásek et al. 2009) (Fig. 1).

The path was found for the holonomic system; further for the non-holonomic system, the path was found to avoid instantaneous turn which is shown in the result.

Then two geodetic coordinates were considered as start and endpoints, and two buildings in between the points are specified as obstacles, using Google Earth (Fig.4). All the geodetic coordinates were converted to the NED (body axis) using the inbuilt function in MATLAB (Mathworks – geodetic2ned 2022; Patrik et al. 2019), and the path was found using RRT and RRT* from starting to ending points (Wang et al. 2020). Then, the waypoints which were taken from the algorithms were again converted to the geodetic frame of reference which is shown in the result (Fig. 2).

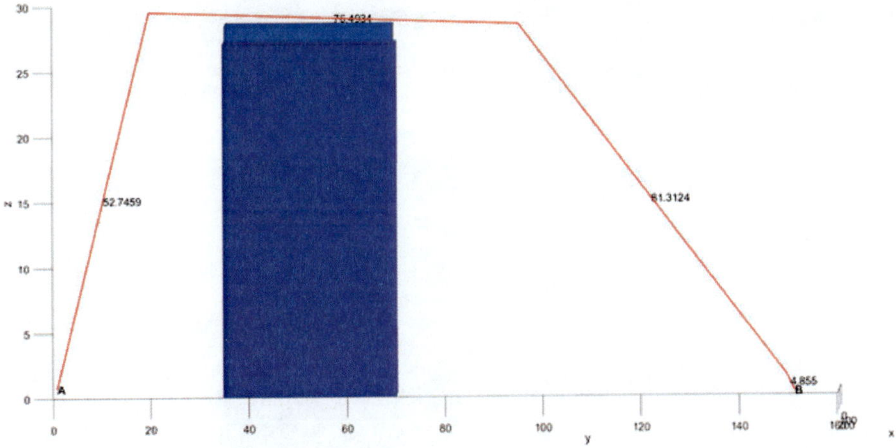

Fig. 1 RRT path for non-holonomic system

Fig. 2 Position marked in Google Earth

3 Conclusion

A random path was generated for RRT and RRT* for a non-holonomic system, and it was plotted in Google Earth for a real-world example, and the distance was found and compared (Figs. 3, 4, 5, 6).

Fig. 3 RRT isometric view

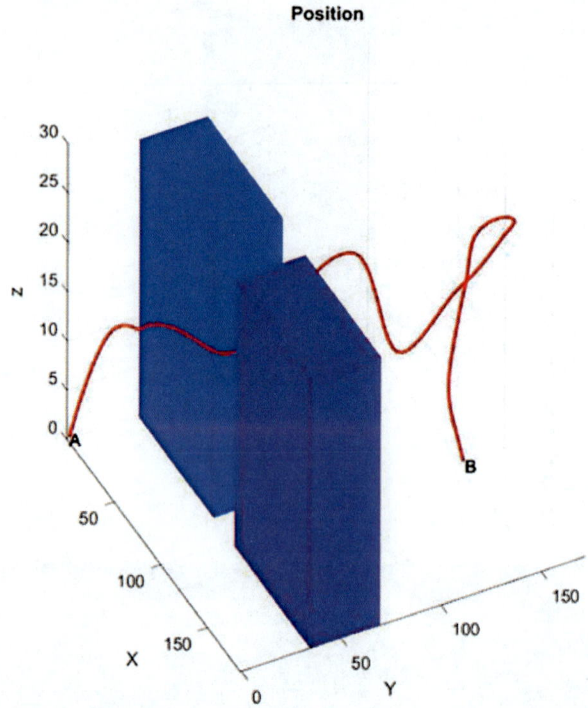

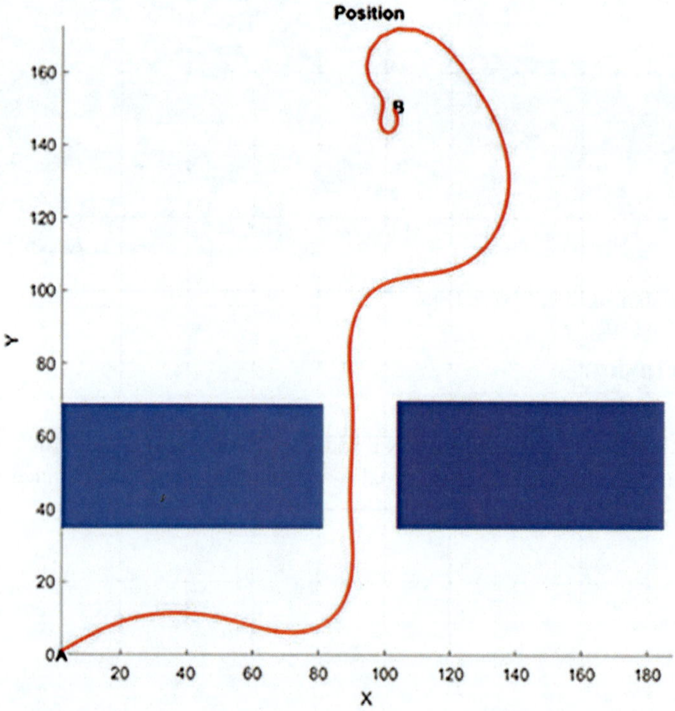

Fig. 4 RRT top view

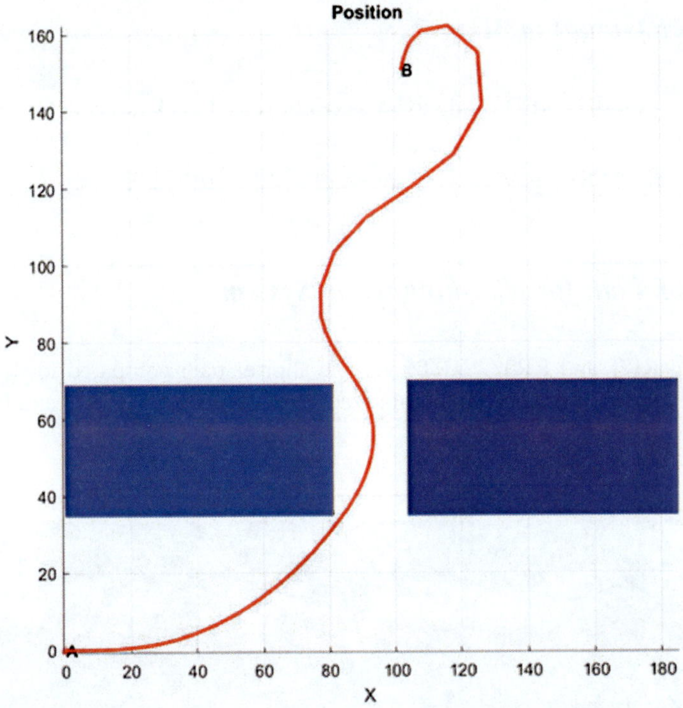

Fig. 5 RRT* Top view

Fig. 6 RRT* Isometric view

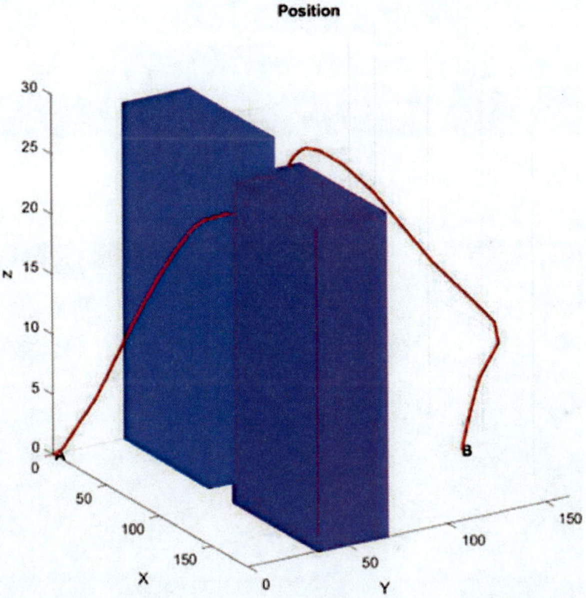

3.1 RRT Output in 3D

Further, the points in NED (body axis) were converted back to geodetic coordinates using the function (ned2geodetic) (Matworks 2022) in MATLAB. Then, the points are plotted in Google Earth by converting them to a KML file.

3.2 RRT Path for Non-holonomic System

Comparing RRT and RRT*, RRT* gives a shorter path compared to RRT which proves the statement RRT* is an optimized version of RRT (Figs. 7 and 8).

Fig. 7 RRT path in Google Earth Distance: – 352 m

Fig. 8 RRT* path in Google Earth Distance: – 281 m

4 Future Work

- All the real-world obstacles are taken as cubes or cuboids in the body frame because the collision-free path of RRT and RRT* 3D only works with cubes or cuboids, extending our work to arbitrary objects.
- Implementing our work on real drones and real-life scenarios.

References

Beard RW, McLain TW (2012) Small unmanned aircraft: theory and practice. Princeton University Press, Princeton

Collinson RP (2013) Introduction to avionics systems. Springer Science & Business Media, New York

Fox D, Burgard W, Thrun S (2002) The dynamic window approach to collision avoidance. IEEE Robot Autom Mag 4:23. https://doi.org/10.1109/100.580977

Gan Y, Zhang B, Ke C, Zhu X, He W, Ihara T (2021) Research on robot motion planning based on RRT algorithm with non- holonomic constraints. Neural Process Lett 53(4):3011–3029. https://doi.org/10.1007/s11063-021-10536-4

Islam F, Nasir J, Malik U, Ayaz Y, Hasan O (2012) RRT*-smart: rapid convergence implementation of RRT* towards optimal solution. In: 2012 IEEE International Conference on Mechatronics and Automation, ICMA 2012. https://doi.org/10.1109/ICMA.2012.6284384

Jha AR (2016) Theory, design, and applications of unmanned aerial vehicles. CRC Press, Boca Raton

Kwak J, Sung Y (2018) Autonomous UAV flight control for GPS-based navigation. IEEE Access 6: 37947–37955. https://doi.org/10.1109/ACCESS.2018.2854712

Lee J, Kwon O, Zhang L, Yoon SE (2012) SR-RRT: selective retraction-based RRT planner. In: Proceedings – IEEE international conference on robotics and automation https://doi.org/10.1109/ICRA.2012.6224928

Mathworks – geodetic2ned (2022), [Online]. Available: https://in.mathworks.com/help/map/ref/geodetic2ned.html. Accessed Mar 2022

Nasir J, Islam F, Malik U, Ayaz Y, Hasan O, Khan M, Muhammad MS (2013) RRT*-SMART: a rapid convergence implementation of RRT*. Int J Adv Robot Syst 10:299. https://doi.org/10.5772/56718

Patrik A, Utama G, Gunawan AAS, Chowanda A, Suroso JS, Shofiyanti R, Budiharto W (2019) GNSS-based navigation systems of autonomous drone for delivering items. J Big Data 6(1): 10.1186/s40537-019-0214- 3

Vonásek V, Faigl J, Krajník T, Přeučil L (2009) RRT-path – a guided rapidly exploring random tree. Lect Notes Control Inf Sci 396. https://doi.org/10.1007/978-1-84882-985-5_28

Wang J, Chi W, Li C, Wang C, Meng MQH (2020) Neural RRT∗: learning-based optimal path planning. IEEE Trans Autom Sci Eng 17(4):1748. https://doi.org/10.1109/TASE.2020.2976560

Structural Synthesis of Euclidean Parallel Robot Manipulators of Spacecraft Docking System

Rasim Ismayil Alizade, Kanan Sabuhi Azimov, and Javad Adalat Samadzade

1 Introduction

Scientific literature contains different descriptions of rectangular parallel extension, retraction, and motion mechanisms of spacecraft docking system (Syromyatnikov 1984; Sean and Scott 2017). Among others, it includes illustrations of octahedral and hexagonal interface Euclidean manipulators of passive and active docking units (Alizade and Samedzade 2021; Alizade and Azimov 2022). Solutions of problems on structural synthesis of spacecraft Euclidean parallel docking manipulators with three, four, five, and six legs were also represented (Alizade 2019). In order to reduce the number of links in the three and four active planes of the support-guide legs, new hexagonal and octahedral interface docking units including new joint "sphere in cylindrical slot" are created.

2 Structural Synthesis of Euclidean Hexagonal Interface Manipulator

Existing spatial mechanisms move in vector space R^3. Consider the new layout of spacecraft and space station docking mechanism by designing Euclidean six-step manipulator in two vector spaces: R^2 and R^3. Thus, combining these spaces led to a new design of the spacecraft docking system. From this point of view, a hexagonal manipulator was considered, which support chains(legs) are at three active planes. The support chains are represented as flat dyads consisting of two links with two rotating pairs. The dyads are then connected to the hexagonal junction (capture ring)

R. I. Alizade · K. S. Azimov · J. A. Samadzade (✉)
National Aviation Academy, Faculty of Air Transport, Department of Flight Apparatus and Aviation Engines, Baku, Azerbaijan

T. H. Karakoc et al. (eds.), *Emerging Trends in Electric Aviation*, Sustainable Aviation, https://doi.org/10.1007/978-3-031-37299-5_15

by means of 4DoF of kinematic pair "sphere-in-cylindrical slot." Using three active faces of the docking unit by means of supporting kinematic chains, a new Euclidean manipulator of the form 6RRSc was obtained.

According to the formula of F. Freudenstain and R.I. Alizade:

$$M = \sum_{i=1}^{j} f_i - \sum_{k=1}^{L} \lambda_k + q_l - j_{ex} \tag{1}$$

where

M – mobility of manipulator.
λ_k – the dimension of the active motion space
L – the number of independent loops
f_i – the DoF of kinetic pairs
j – the number of joints,
q_l – the excessive over closing constraints
j_{ex} – the number of passive DoF in kinematic pairs.

Euclidean docking mechanism with $q_l = 0$; $j_{ex} = 0$; $\lambda_k = 6$; $L = 5$; $\sum_{i=1}^{j} f_i = 36$ is represented (Fig. 1).

Using a formula and input data, the mobility of manipulator is equal

$$M = 36 - 6 \cdot 5 - 0 - 0 = 6$$

Consider another version of the hexagonal manipulator with three legs of 3(5R) Scs structure which connect the odd sides of hexagonal capture ring and docking

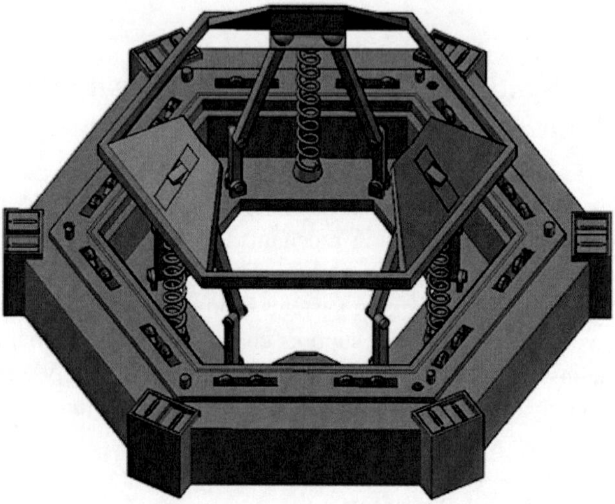

Fig. 1 New design of hexagonal spacecraft docking system

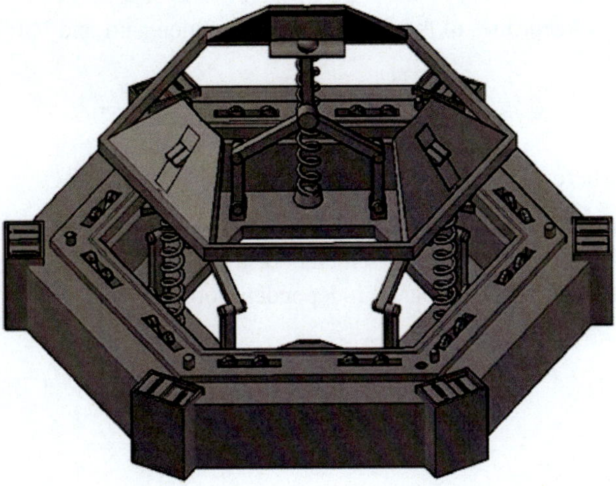

Fig. 2 The 6DoF parallel soft capture ring of spacecraft docking unit

unit. Closed kinematic circuits 5R create flat five-link mechanism with two degrees of freedom M = 2, the connecting coupler point of which is connected to the hinge Scs. Thus, the advantage is that the combination of planar structures with spatial structures is used here for the first time.

Parallel Euclidean docking mechanism with $\lambda = 6$, $\lambda_1 = 3$, $c_1 = 3$, M = 6 is represented (Fig. 2). The end effector motion parameters: $\lambda = 6$; the loop motion of the legs: $\lambda_1 = 3$; the total number of legs: $sc_1 = 3$; mobility of parallel docking mechanism: M = 6.

By using Eq. (1), total DoF and kind of kinematic pairs at the three legs can be calculated as

$$\sum\nolimits_{k=1}^{5} \lambda_k = 3 \cdot 3 + 6 \cdot 2 = 21; \sum f_i = 5 \cdot 3 + 4 \cdot 3 = 27; M = 27 - 21 - 0 - 0 = 6$$

3 Structural Synthesis of Euclidean Octahedral Manipulator

The proposed new construction of the spacecraft docking unit consists of an octahedral manipulator supported by chains(legs) which move in four different active planes. The supporting chains are represented as planar dyads consisting of two links with two revolute pairs and by higher pair connected to docking system. The dyads are then connected to the octagonal junction (capture ring) by means of 4DoF kinematic pairs "sphere in cylindrical slot." Using the four active faces of the docking module by means of supporting kinematic chains, a new Euclidean manipulator of the form 4RRScs is obtained.

According to the formula of F. Freudenstain and R.I. Alizade:

$$M = \sum\nolimits_{i=1}^{j} f_i - \lambda L \tag{2}$$

where

M = 6 – is the mobility of manipulator.
λ = 6 – the motion parameters of independent loops,
L = 3 – the number of independent loops

$$6 = \sum\nolimits_{i=1}^{j} f_i - 6 \cdot 3; \sum\nolimits_{i=1}^{j} f_i = 24$$

So the sum of degrees of freedom of kinematic pairs 4RRScs, i.e., $\sum_{i=1}^{j} f_i = 24$.

Total number of degrees of freedom of the kinematic pairs on one support of the kinematic chain (Fig. 3):

$$j_l = \frac{\sum_{i=1}^{j} f_i}{C_l} = \frac{24}{4} = 6$$

Thus, the advantage is that here, for the first time, the combination of four planar structures with spatial ones is applied. The platform itself, which connects with the space station platform, makes six movements in space, both translational along X, Y,

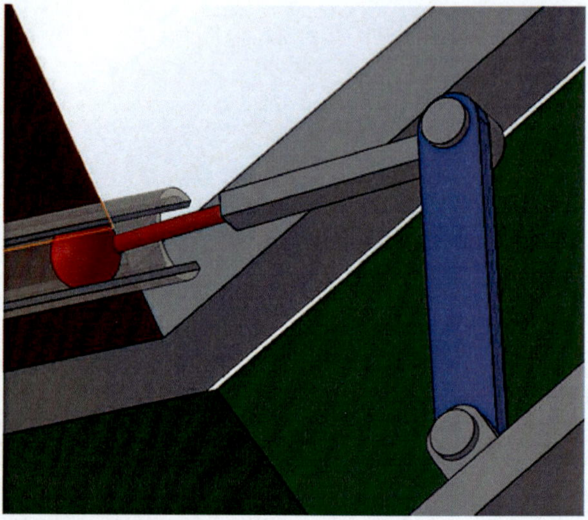

Fig. 3 Dyad connection with an octahedral coupling element(ring) with a sphere-in-cylinder kinematic pair having four degrees of freedom

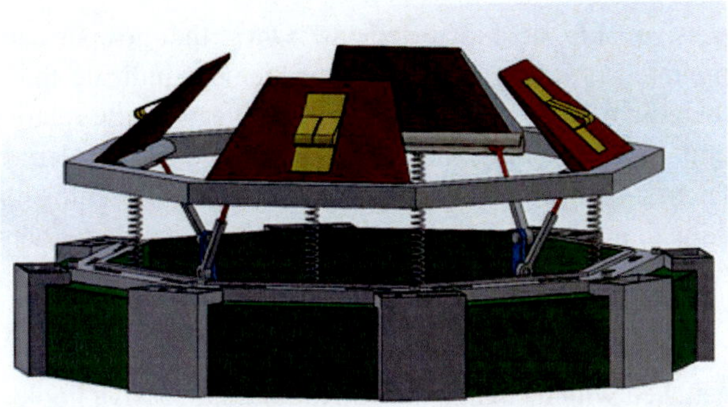

Fig. 4 New design of the octahedral spacecraft docking manipulator

and Z parameters and rotational around X, Y and Z axes, showing the orientation of the lock. Consequently, six parameters quite allow us to perform the process of docking and orientation of the spacecraft with the space station at low speeds. As a result, after conducting axial orientation, the docking manipulator of the spacecraft slowly docks with the space station. The corresponding locks being actuated and resulting in the connection of the spacecraft with the space station. After connection airtight, a cosmonaut is able to transit through this manipulator from the spacecraft to the space station.

It should also be noted that the possibility of designing four supporting kinematic chains instead of six reduces the weight of the structure, which is one of the main requirement for docking devices of spacecrafts (Fig. 4).

The arrangement of the dyads of the docking mechanism is such that the octagonal capture ring is aligned with the tunnel. The control architecture uses this fact to simplify the control of actuators during soft capture and alignment of the docking element.

The problem of structural synthesis of Euclidean docking mechanism with variable general constraint of the legs and closed loops depends on the DoF and motion of the manipulator. Mobility equations were introduced for manipulators with hexagonal and octahedral interface. The motion of end effector legs with new joint "sphere in cylindrical slot" at different Euclidean planes was considered.

During approach, the soft-grip system advances to the "ready to grab" position, where it waits for initial contact between the two spacecraft. The spacecraft guidance, navigation, and control system ensure that the initial contact occurs within the docking system's capture area.

Upon contact, forces between the rough alignment guides cause the dyads to shift from the ready position to the capture position. This event signals the controller to begin a maneuver known as "Lunge." During the dropout, the soft grip system expands appropriately to compensate for spacecraft misalignments and engages the soft grip latches mounted on each of the three guide lobes (petals) of the docking element. Each latch is in contact with the opposite locking plate mounted behind the passive side docking element. Once the gripper latches are engaged, sensors mounted on the gripper docking element indicate that the grip has been reached, thereby signaling the system to begin easing the relative motion of the two spacecraft. At this point, the two hexagonal docking elements are held loosely by the latches. As soon as the time allotted for loosening expires, the controller starts adjusting the lengths of the actuators to align them, thereby aligning the two mating tunnels. When the alignment is complete, the hexagonal docking elements retract, closing the gap between the two tunnels and engaging their shear elements.

When the two tunnels are in close proximity, a set of 16 structural hook locks are actuated which engage the corresponding passive hooks on the opposite tunnel. The structural hooks close the remaining gap between the tunnels, compressing the elastomeric seal and creating a tight connection. Once the hard grip is complete, the motorized connectors engage, allowing energy and data to be transferred between the docked spacecraft.

References

Alizade R (2019) Structural synthesis of robot manipulators by using screw with variable pitch. Univers J Mech Eng 7(2):50–63

Alizade RI, Azimov KS (2022) Initial conditions for spacecraft docking. Readings:54–56

Alizade RI, Samedzade JA (2021) A hexagonal interface spacecraft docking system. Robot Autom Eng J 5(2):1–3

Sean MK, Scott PC (2017) International Docking Standard (IDSS) Interface Definition Document (IDD)

Syromyatnikov V.S (1984) Docking devices of spacecraft. In: Mechanical Engineering, pp 216

Future Prospects for Fuel-Cell Aircraft: Challenges and Opportunities

Anita Prapotnik Brdnik and Maršenka Marksel

Nomenclature

ICE	Internal combustion engine
MTOM	Maximum take-off mass

1 Introduction

The invention of the internal combustion engine (ICE) in the nineteenth century ushered in the technological revolution and improved our quality of life. Nowadays, however, there is a growing awareness of the negative consequences of the ICE discovery and use of fossil fuels. The toxic gasses emitted, such as NO_X, SO_X, CO, and hydrocarbons, as well as noise, can cause various diseases, including heart, lung, and liver diseases, as well as stress symptoms (Brasseur et al. 2016). Although aviation today accounts for only 2% of global greenhouse gas emissions and 3% in Europe, it has increased by 70% between 2005 and 2020 and will continue to grow

A. P. Brdnik (✉)
Faculty of Civil Engineering, Transportation Engineering and Architecture, University of Maribor, Maribor, Slovenia

Faculty of Energy Technology, University of Maribor, Krško, Slovenia
e-mail: anita.prapotnik@um.si

M. Marksel
Faculty of Civil Engineering, Transportation Engineering and Architecture, University of Maribor, Maribor, Slovenia
e-mail: marsenka.marksel@um.si

T. H. Karakoc et al. (eds.), *Emerging Trends in Electric Aviation*, Sustainable Aviation, https://doi.org/10.1007/978-3-031-37299-5_16

in the future (Owen et al. 2010; European Commision 2020). Several studies forecast annual growth of 4.5% to 4.8% (Airbus 2015; Boeing 2014), which could lead to a 300% increase in CO_2 emissions by 2050 (European Commision 2020). In addition, emissions of greenhouse gasses such as CO_2 and N_20 can have lasting and devastating effects on the climate and environment. Therefore, numerous new studies and projects funded by the world's governments are trying to find suitable alternatives to fossil fuels and new green ways of producing energy.

As a result, new fuel and propulsion technologies are being proposed, including battery-powered aircraft, hydrogen-powered aircraft, hybrid systems, and the use of synthetic fuels. Each of these proposals has its advantages and disadvantages compared to the ICE technology. The suitability of these solutions may vary from one application to another, and the solution that is suitable for ground transportation problems may be unsuitable for aircraft due to different specifications. Even more, the best solution for small general aviation aircraft may not be suitable for large long-range aircraft and vice versa. Therefore, different scenarios for different applications must be tested, and the best solution for each application must be found.

Aircraft can use hydrogen as a fuel either by burning it in a manner similar to kerosene or by using it in fuel cells to generate electricity that drives the propeller through an electric motor. In the first case, the aircraft still emits NO_X gasses, albeit in smaller quantities than with a conventional ICE engine (Khandelwal et al. 2013). In the second case, the only byproduct is water vapor. Although water vapor is also considered as greenhouse gas at high altitudes, this is not considered a problem at the altitudes where propeller planes fly. In addition, hydrogen can be produced by various processes that may be more or less environmentally friendly. The cheapest and most common method of producing hydrogen today is steam reforming (Adolf et al. 2017). This still emits CO2, albeit in smaller quantities than kerosene combustion. In the future, however, it is planned to produce hydrogen either by steam reforming with carbon recovery (blue hydrogen) or by electrolysis with electricity from renewable energy sources (green hydrogen) or from nuclear power plants (red hydrogen) (IRENA 2020).

2 Technical Challenges and Opportunities of Fuel Cell Aircraft

Although all the components for the fuel cell aircraft propulsion system have been developed, they still need to be adapted for use in the aircraft. The biggest challenge is to achieve a sufficiently low MTOM, since the aircraft's fuel consumption is linearly proportional to its mass. Studies show that the 11-seat fuel cell aircraft would either suffer significant losses in mission performance compared to conventional aircraft or have a higher MTOM than conventional aircraft (Trainelli et al. 2019). However, depending on the future development of fuel cell power-to-weight ratio, the MTOM of 19-seat aircraft may be similar to the MTOM of 19-seat conventional aircraft in the optimistic case scenario (Vonhoff 2021; Marksel and Prapotnik Brdnik 2022).

The cryogenic and pressurized tanks intended for use in ground transportation (trucks, trains) are too heavy for use in aircraft. Therefore, new tanks must be designed and manufactured to meet aircraft needs and specifications. Several theoretical studies have been conducted on the composition, shape, and location of such tanks. Studies (Verstraete et al. 2010; Winnefeld et al. 2018) propose a single cylindrical cryogenic hydrogen tank integrated into the fuselage with a circular cross-section, while the (Silberhorn et al. 2022) study proposes an arrangement of the tanks in pods. In addition, both (Verstraete et al. 2010; Winnefeld et al. 2018) predict that the cryogenic hydrogen tank for aircraft can achieve gravimetric densities between 0.6 and 0.7. Development of one of the first flight-qualified tank is being conducted by Gloyer-Taylor Laboratories (GTL). The 2.4 m long tank with a diameter of 1.2 m and a mass of 67 kg has the ability to store 120 kg of hydrogen (gravimetric ratio of 0.7) (Blain 2022).

Although the power-to-weight ratio of the electric motor can be significantly higher than the power-to-weight ratio of ICE, the fuel cells themselves have a low power-to-weight ratio. Fig. 1. shows the MTOM dependence of a 19-seat aircraft on power-to-weight ratio of fuel cells. Nowadays, fuel cells alone reach power to weigh rations of around 2–3 kW/kg. It is evident from the figure that this value has to be improved in the future (Marksel and Prapotnik Brdnik 2022).

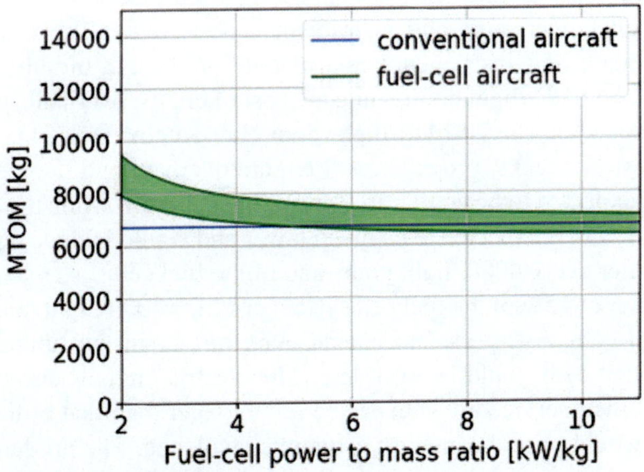

Fig. 1 Dependence of the MTOM of 19-seat aircraft on the power to weight ratio of fuel cells. (Marksel and Prapotnik Brdnik 2022)

3 Economical Challenges and Opportunities

An important aspect to consider in the future of fuel cell aircraft is economic justification, in which operating costs will play a critical role. Therefore, the direct operating costs of the aircraft, which consist of depreciation, interest, insurance, fuel, maintenance, crew costs, and fees (airport, navigation), must be reviewed and compared to those of a conventional reference aircraft.

The cost of depreciation, interest, and insurance is proportional to the price of the aircraft. The difference between the price of fuel cell aircraft and comparable conventional aircraft depends on the price of the fuel cell system, electric motor system, cryogenic tank, and MTOM. It is predicted that the price of fuel cell systems will decrease in the future reaching values of 40 EUR/kW in Europe (FCH 2018) and 30 dollars/kW in the USA (Thompson et al. 2018). Electric motor price can be estimated to 100–150 EUR/kW (Finger et al. 2019), while the cryotank price is expected to reach a value of 400 EUR/kg by 2050 (McKinsey and Company 2020; Mukhopadhaya and Rutherford 2022). The cost of other components is expected to either remain the same as for conventional aircraft or scale with MTOM. The engine system of fuel cell aircraft is therefore expected to be cheaper than the engine system of conventional aircraft (about 800 EUR/kW), while the price of the entire aircraft will depend on the MTOM and will probably be slightly higher than the price of conventional aircraft.

Maintenance costs include airframe and engine checks, engine overhauls, oil changes, propeller overhauls, and instrument replacements. The cost of checks, propeller overhaul, and instrument replacement for the Let-turbolet L410NG is estimated at 90.56 EUR/flight hour, while the cost of engine overhaul and oil change for the is estimated at 162.70 EUR/flight hour (Let Kunovice 2021). For fuel cell aircraft, the cost of checks, propeller, and engine overhaul and the replacement of instruments is expected to be similar to convectional aircraft, while the maintenance of the engine system is expected to be much lower and would include replacement of the fuel cells after every 4000 flight hours and of the fuel cell system components at an estimated cost of 15% of the fuel cell system price for 8000 flight hours. Since the life of the fuel cells is expected to extend even further in the future, the cost of replacing the fuel cells could be even less. The electric engines do not need to be overhauled, as their service life is expected to be longer than that of the aircraft.

Fuel costs would pose the greatest economic challenge. The production price of green hydrogen is expected to reach values of about 3 EUR/kg by 2030 (STATISTA 2021), while the costs of liquefaction, distribution, and sales could double this price (Connelly et al. 2019). Due to the higher energy density of hydrogen (33 kWh/kg compared to 12 kWh/kg of kerosene) and the higher efficiency of the fuel cell system (about 50\% compared to about 33\% for turboprop engines), the fuel cell aircraft is expected to consume between 3.5 and 4.2 less fuel in kg than a convectional engine aircraft, which means that the price of hydrogen would still be too high unless the price of kerosene continues to rise due to environmental taxes.

4 Conclusions

In addition to the technical and economic challenges and opportunities, other factors may influence the acceptance of fuel cell aircraft, including safety considerations, regulatory issues, ground infrastructure, and government and public acceptance. Hydrogen is a hazardous and highly flammable substance, but we have now learned to deal with it. The ground infrastructure for refueling aircraft with hydrogen is not yet in place. However, in the initial phase, aircraft refueling can be done directly from a mobile cryogenic truck, and no special investment in ground infrastructure is required except for connection and safety equipment. Public acceptance may be positive due to environmental issues or negative due to safety issues and unknown new technologies. The role of political support and will of governments could be critical and include the provision of regulatory frameworks, support for research and innovation, necessary investments in airport infrastructure and aircraft development, environmental fees, and fee deductions for environmentally friendly technologies. All in all, with further development, fuel cell aircraft represent a viable and environmentally friendly alternative to conventional aircraft.

References

Adolf J, Balzer C, Louis J, Schabla U (2017) Shell hydrogen study: energy of the future; technical report. Shell Deuchland Oil GmbH, Hamburg

Airbus (2015) Global Market Forecas7: Flying By Numbers 2015–2034. https://espas.secure.europarl.europa.eu/orbis/sites/default/files/generated/document/en/Global_Market_Forecast_2015_Book.pdf

Blain L (2022) Ultra-light liquid hydrogen tanks promise to make jet fuel obsolete. https://newatlas.com/aircraft/hypoint-gtl-lightweight-liquid-hydrogen-tank/

Boeing Commercial Airplanes (2014) Current Marker Outlook 2014–2033. http://i2.cdn.turner.com/cnn/2015/images/09/02/boeing_current_market_outlook_2014.pdf

Brasseur GP et al (2016) Impact of aviation on climate: FAA's aviation climate change research initiative (ACCRI) phase II. Bull Am Meteorol Soc 97:561–583. https://doi.org/10.1175/BAMS-D-13-00089.1

Connelly E, Penev M, Elgowainy A, Hunter C, (2019) Current status of hydrogen liquefaction costs, doe hydrogen and fuel cells program record. https://www.hydrogen.energy.gov/pdfs/19001_hydrogen_liquefaction_costs.pdf

European Commission (2020) Sustainable and smart mobility strategy – putting European transport on track for the future. https://eur-lex.europa.eu/legal-content/EN/TXT/?uri=COM:2020:789:FIN

FCH 2 JU (2018) Multi – annual work plan 2014–2020. https://www.fch.europa.eu/sites/default/files/MAWP%20final%20version_endorsed%20GB%2015062018%20%28ID%203712421%29.pdf

Finger DF, Goetten F, Braun C, Bil C (2019) Cost estimation methods for hybrid-electric general aviation aircraft. In Asia-pacifc international symposium on aerospace technology. https://www.researchgate.net/profile/Falk-Goetten/publication/337757069_Cost_Estimation_Methods_for_Hybrid-Electric_General_Aviation_Aircraft/links/5de87703299bf10bc3405695/Cost-Estimation-Methods-for-Hybrid-Electric-General-Aviation-Aircraft.pdf

International Renewable Energy Agency (2020) Green hydrogen cost reduction. https://www.irena.org/publications/2020/Dec/Green-hydrogen-cost-reduction

Khandelwal B, Karakurt A, Sekaran PR, Sethi V, Singh R (2013) Hydrogen powered aircraft: the future of air transport. Prog Aerosp Sci 60:45–59. https://doi.org/10.1016/j.paerosci.2012.12.002

Let Kunovice (2021) DOC Data sheet for Let Turbolet 410 NG obtained directly from a manufacturer

Marksel M, Prapotnik Brdnik A (2022) Maximum take-off mass estimation of a 19-seat fuel cell aircraft consuming liquid hydrogen. Sustainability 14(14):8392. https://doi.org/10.3390/su14148392

McKinsey & Company (2020) Hydrogen-powered aviation a fact-based study of hydrogen technology, economics, and climate impact by 2050. The Print Agency, Ixelles. https://doi.org/10.2843/766989

Mukhopadhaya J, Rutherford D (2022) Performance analysis of evolutionary hydrogen-powered aircraft. International council on clean transportation, Washington. https://theicct.org/wpcontent/uploads/2022/01/LH2-aircraft-white-paper-A4-v4.pdf

Owen B, Lee DS, Lim L (2010) Flying into the future: aviation emissions scenarios to 2050. Environ Sci Technol 44(7):2255–2260. https://doi.org/10.1021/es902530z

Silberhorn D, Atanasov G, Walther JN, Zill T (2022) Assesment of hydrogen fuel tank integration at aircraft level. https://core.ac.uk/ download/pdf/237080603.pdf

STATISTA (2021) Production costs of green hydrogen worldwide by select country in 2020, with a forecast until 2050. https://www-statista-com.ezproxy.lib.ukm.si/statistics/1086695/green-hydrogen-cost-development-by-country/

Thompson ST et al (2018) Direct hydrogen fuel cell electric vehicle cost analysis: system and high-volume manufacturing description, validation, and outlook. J Power Sources 399(30):304–313. https://doi.org/10.1016/j.jpowsour.2018.07.100

Trainelli L, Salucci F, Comincini D, Riboldi CED, Rolando A (2019) Sizing and performance of hydrogen-driven airplanes. Italian Association of Aeronautic and Astronautic, XXV International Congress Italy. https://mahepa.eu/wp-content/uploads/2020/06/Sizing-and-Performance-of-Hydrogen-Driven-Airplanes.pdf

Verstraete D, Hendrick P, Pilidis P, Ramsden K (2010) Hydrogen fuel tanks for subsonic transport aircraft. Int J Hydrog Energy 35(20):11085–11098. https://doi.org/10.1016/j.ijhydene.2010.06.060

Vonhoff, GLM (2021) Conceptual desigh of fuel cell aircraft. PhD thesis. https://repository.tudelft.nl/islandora/object/uuid\%3A8bd63dec-b67b-496b-92bc-3d5c07ff859f

Winnefeld C, Kadyk T, BensmannB KU, Hanke-Rauschenbach R (2018) Modelling and designing cryogenic hydrogen tanks for future aircraft applications. Energies 11(1):105. https://doi.org/10.3390/en11010105

Index

T. H. Karakoc et al. (eds.), *Emerging Trends in Electric Aviation*, Sustainable Aviation, https://doi.org/10.1007/978-3-031-37299-5